m 770
9

Universitext

Universitext

Editors (North America): J.H. Ewing, F.W. Gehring, and P.R. Halmos

Aupetit: A Primer on Spectral Theory
Aksoy/Khamsi: Nonstandard Methods in Fixed Point Theory
Aupetit: A Primer on Spectral Theory
Berger: Geometry I, II (two volumes)
Bliedtner/Hansen: Potential Theory
Bloss/Bleeker: Topology and Analysis
Cecil: Lie Sphere Geometry: With Applications to Submanifolds
Chandrasekharan: Classical Fourier Transforms
Charlap: Bierbach Groups and Flat Manifolds
Chern: Complex Manifolds Without Potential Theory
Cohn: A Classical Invitation to Algebraic Numbers and Class Fields
Curtis: Abstract Linear Algebra
Curtis: Matrix Groups
van Dalen: Logic and Structure
Devlin: Fundamentals of Contemporary Set Theory
Edwards: A Formal Background to Mathematics I a/b
Edwards: A Formal Background to Mathematics II a/b
Foulds: Graph Theory Applications
Emery: Stochastic Calculus
Fukhs/Rokhlin: Beginner's Course in Topology
Frauenthal: Mathematical Modeling in Epidemiology
Gallot/Hulin/Lafontaine: Riemannian Geometry
Gardiner: A First Course in Group Theory
Gårding/Tambour: Algebra for Computer Science
Godbillon: Dynamical Systems on Surfaces
Goldblatt: Orthogonality and Spacetime Geometry
Humi/Miller: Second Course in Order Ordinary Differential Equations
Hurwitz/Kritkos: Lectures on Number Theory
Jones/Morris/Pearson: Abstract Algebra and Famous Impossibilities
Iverson: Cohomology of Sheaves
Kelly/Matthews: The Non-Euclidean Hyperbolic Plain
Kempf: Complex Abelian Varieties and Theta Functions
Kostrikin: Introduction to Algebra
Krasnoselskii/Pekrovskii: Systems with Hypstersis
Luecking/Rubel: Complex Analysis: A Functional Analysis Approach
MacLane/Moerdijk: Sheaves in Geometry and Logic
Marcus: Number Fields
McCarthy: Introduction to Arithmetical Functions
Meyer: Essential Mathematics for Applied Fields
Mines/Richman/Ruitenburg: A Course in Constructive Algebra
Moise: Introductory Problem Course in Analysis and Topology
Montesinos: Classical Tessellations and Three Manifolds
Nikulin/Shafarevich: Geometries and Group
Øskendal: Stochastic Differential Equations
Rees: Notes on Geometry

(continued after index)

Thomas E. Cecil

Lie Sphere Geometry

With Applications to Submanifolds

With 14 Illustrations

Springer-Verlag

New York Berlin Heidelberg London Paris
Tokyo Hong Kong Barcelona Budapest

Thomas E. Cecil
Department of Mathematics
College of the Holy Cross
Worcester, MA 01610
USA

Editorial Board
(North America):

J.H. Ewing
Department of Mathematics
Indiana University
Bloomington, IN 47405
USA

F.W. Gehring
Department of Mathematics
University of Michigan
Ann Arbor, MI 48109
USA

P.R. Halmos
Department of Mathematics
Santa Clara University
Santa Clara, CA 95053
USA

Mathematics Subject Classifications (1980): 53C42, 53A07, 53A20

Library of Congress Cataloging-in-Publication Data
Cecil, T. E. (Thomas E.)
    Lie sphere geometry: with application to submanifolds/Thomas
E. Cecil.
        p. cm. — (Universitext)
    Includes bibliographical references and index.
    ISBN 0-387-97747-3. — ISBN 3-540-97747-3
    1. Geometry, Differential. 2. Submanifolds. I. Title.
II. Series.
QA649.C42   1992
516.3′6 — dc20
                                                    91-4095

Printed on acid-free paper.

Camera-ready copy prepared by the author.
Printed and bound by R.R. Donnelley & Sons, Harrisonburg, VA.
Printed in the United States of America.

9 8 7 6 5 4 3 2 1

ISBN 0-387-97747-3 Springer-Verlag New York Berlin Heidelberg
ISBN 3-540-97747-3 Springer-Verlag Berlin Heidelberg New York

To my sons,

Tom, Mark and Michael

Preface

The purpose of this monograph is to provide an introduction to Lie's geometry of oriented spheres and its recent applications to the study of submanifolds of Euclidean space. Lie [1] introduced his sphere geometry in his dissertation, published as a paper in 1872, and used it in his study of contact transformations. The subject was actively pursued through the early part of the twentieth century, culminating with the publication in 1929 of the third volume of Blaschke's [1] *Vorlesungen über Differentialgeometrie*, which is devoted entirely to Lie sphere geometry and its subgeometries. After this, the subject fell out of favor until 1981, when Pinkall [1] used it as the principal tool in his classification of Dupin hypersurfaces in $\mathbb{R}^4$. Since that time, it has been employed by several geometers in the study of Dupin, isoparametric and taut submanifolds.

This book is not intended to replace Blaschke's work, which contains a wealth of material, particularly in dimensions two and three. Rather, it is meant to be a relatively brief introduction to the subject, which leads the reader to the frontiers of current research in this part of submanifold theory. Chapters 1 and 2 are accessible to a beginning graduate student who has taken courses in linear and abstract algebra and projective geometry. Chapters 3 and 4 contain the applications to submanifold theory. These chapters require a first graduate course in differential geometry as a necessary background. A detailed description of the contents of the individual chapters is given in the Introduction, which also serves as a survey of the field to this point in time.

I wish to acknowledge certain works which have been especially useful to me in writing this book. Much of Chapters 1 and 2 is based on Blaschke's book. The proof of the Cartan–Dieudonné theorem in Section 2.2 is taken from E. Artin's book [1], *Geometric Algebra*. Two sources are particularly influential in Chapters 3 and 4. The first is Pinkall's dissertation [1] and his subsequent paper [4], which have proven to be remarkably fruitful. Secondly, the approach to the study of Legendre submanifolds using the method of moving frames is due to Shiing–Shen Chern, and was first presented in two papers by Chern and myself [1]–[2]. These two papers and indeed this monograph grew out of my work with Professor Chern during my 1985–86 sabbatical at Berkeley. I am very grateful to Professor Chern for many helpful discussions and insights.

I also want to thank several other mathematicians for their personal contributions. Katsumi Nomizu introduced me to Pinkall's work and Lie sphere geometry in 1982, and his seminar at Brown University has been the site of many enlightening discussions on the subject since that time. Thomas Banchoff introduced me to the cyclides of Dupin in the early seventies, when I was a graduate student, and he has provided me with several key insights over the years, particularly through his films. Patrick Ryan has contributed significantly to my understanding of this subject through many lectures and discussions. I also want to acknowledge helpful conversations and correspondence on various aspects of the subject with Steven Buyske, Sheila Carter, Leslie Coghlan, Josef Dorfmeister, Thomas Hawkins, Wu–Yi Hsiang, Nicolaas Kuiper, Martin Magid, Reiko Miyaoka, Ross Niebergall, Tetsuya Ozawa, Richard Palais, Ulrich Pinkall, Helmut Reckziegel, Chuu–Lian Terng, Gudlaugur Thorbergsson, and Alan West.

This book grew out of lectures given in the Brown University Differential Geometry Seminar in 1982–83 and subsequent lectures given to the Clavius Group during the summers of 1985–89 at the University of Notre Dame, the University of California at Berkeley, Fairfield University and the Institute of Advanced Study. I want to thank my fellow members of the Clavius Group for their support of these lectures and many enlightening remarks. I also acknowledge with gratitude the hospitality of the institutions mentioned above.

I wish to thank my colleagues in the Department of Mathematics at the College of the Holy Cross, several of whom are my former teachers, for many insights and much encouragement over the years. I especially wish to mention my first teacher in linear algebra and real analysis, Leonard Sulski, who recently passed away after a courageous battle against leukemia. Professor Sulski was a superb, dedicated teacher, and a good and generous man. He will be missed by all who knew him.

While writing this book, I was supported by grants from the National Science Foundation (DMS–8907366 and DMS–9101961) and by a Faculty Fellowship from the College of the Holy Cross. This support is gratefully acknowledged.

I want to thank my three undergraduate research assistants from Holy Cross, Michele Intermont, Christopher Butler and Karen Purtell, who were also supported by the NSF. They worked through various versions of the manuscript and made many helpful comments. I also wish to thank the mathematics editorial department of Springer–Verlag for their timely professional help in preparing this manuscript for publication, and Kenneth Scott of Holy Cross for his assistance with the word–processing program.

Finally, I am most grateful to my wife, Patsy, and my sons, Tom, Mark and Michael, for their patience, understanding and encouragement during this lengthy project.

Thomas E. Cecil
College of the Holy Cross
Worcester, Massachusetts

August, 1991

Contents

Introduction

Lie [1] introduced his geometry of oriented spheres in his dissertation, published as a paper in *Mathematische Annalen* in 1872. Sphere geometry was also prominent in his study of contact transformations (Lie–Scheffers [1]) and in Volume III of Blaschke's [1] *Vorlesungen über Differentialgeometrie*, published in 1929. In recent years, sphere geometry has become a valuable tool in the study of Dupin submanifolds in Euclidean space $\mathbb{R}^n$, beginning with Pinkall's [1] dissertation in 1981. In this introduction, we will outline the contents of the book and mention some related results.

Lie established a bijective correspondence between the set of all oriented hyperspheres, oriented hyperplanes and point spheres in $\mathbb{R}^n \cup \{\infty\}$ and the set of points on the quadric hypersurface Q^{n+1} in real projective space $\mathbb{P}^{n+2}$ given by the equation $< x, x > = 0$, where $< , >$ is an indefinite scalar product with signature $(n+1, 2)$ on $\mathbb{R}^{n+3}$. The quadric Q^{n+1} contains projective lines but no linear subspaces of higher dimension. The one–parameter family of oriented spheres in $\mathbb{R}^n$ corresponding to the points on a line on Q^{n+1} is called a *parabolic pencil* of spheres. It consists of all oriented hyperspheres in oriented contact at a certain contact element on $\mathbb{R}^n$. In this way, Lie also established a bijective correspondence between the manifold of contact elements and the manifold Λ^{2n-1} of projective lines on Q^{n+1}. The details of these considerations are given in Chapter 1.

A *Lie sphere transformation* is a projective transformation of $\mathbb{P}^{n+2}$ which maps Q^{n+1} to itself. In terms of the geometry of $\mathbb{R}^n$, a Lie sphere transformation maps oriented spheres to oriented spheres. Furthermore, since a projective transformation maps lines to lines, a Lie sphere transformation preserves oriented contact of spheres in $\mathbb{R}^n$. Lie proved the so–called

"Fundamental Theorem of Lie sphere geometry" in the case $n = 3$, and Pinkall
[4] generalized this to arbitrary dimension. The theorem states that any
line–preserving diffeomorphism of Q^{n+1} is the restriction to Q^{n+1} of a
projective transformation of $\mathbb{P}^{n+2}$. In other words, a transformation on the
space of oriented spheres which preserves oriented contact is a Lie sphere
transformation. One can show that a Lie sphere transformation is induced by
an orthogonal transformation of $\mathbb{R}^{n+3}$ endowed with the metric $< , >$. Thus, the
group G of Lie sphere transformations is isomorphic to the quotient group
$O(n+1,2)/\{\pm I\}$. By the theorem of Cartan and Dieudonné, the orthogonal
group $O(n+1,2)$ is generated by inversions in hyperplanes, and therefore so is
G. Any Moebius (conformal) transformation of $\mathbb{R}^n \cup \{\infty\}$ induces a Lie sphere
transformation, and the Moebius group is precisely the subgroup of Lie sphere
transformations which map point spheres to point spheres. In Chapter 2, we
prove these results and give a geometric description of inversions. We also
discuss the sphere geometries of Laguerre and Moebius. These, as well as the
usual Euclidean, spherical and hyperbolic geometries, are subgeometries of Lie
sphere geometry.

The manifold Λ^{2n-1} of lines on the quadric Q^{n+1} has a contact structure,
i.e., a 1–form ω such that $\omega \wedge (d\omega)^{n-1}$ does not vanish on Λ^{2n-1}. The condition
$\omega = 0$ defines a codimension one distribution D on Λ^{2n-1} which has integral
submanifolds of dimension $n-1$ but none of higher dimension. An immersion
$\lambda : M \to \Lambda^{2n-1}$ of an $(n-1)$–dimensional manifold such that $\lambda^* \omega = 0$ is called a
Legendre submanifold. These are studied in detail in Chapter 3.

A hypersurface M in $\mathbb{R}^n$ naturally induces a Legendre submanifold.
More generally, an immersed submanifold V of codimension greater than one
in $\mathbb{R}^n$ induces a Legendre submanifold whose domain is the unit normal bundle
B^{n-1} of V in $\mathbb{R}^n$. Thus, Lie sphere geometry can be used to study any problem
concerning submanifolds of $\mathbb{R}^n$, or more generally of the sphere S^n or
hyperbolic space H^n. Of course, Lie sphere geometry is particularly
well–suited for the study of problems which deal with spheres in some way.
A large class of such problems are those involving the principal curvatures of
a submanifold, since each principal curvature gives rise to a corresponding
curvature sphere.

Let M be a hypersurface in a real space–form $\mathbb{R}^n$, S^n or H^n. The eigenvalues of the shape operator A of M are called *principal curvatures*, and their corresponding eigenspaces are called *principal spaces*. A submanifold S of M is called a *curvature surface* if at each point x of S, the tangent space $T_x S$ is a principal space. This generalizes the classical notion of a line of curvature of a surface in 3–space. Curvature surfaces are abundant, for there always exists an open dense subset Ω of M on which the multiplicities of the principal curvatures are locally constant (See Reckziegel [1]–[2]). If a principal curvature κ has constant multiplicity m on some open set $U \subset M$, then the corresponding distribution of principal spaces is an m–dimensional foliation, and the leaves of this *principal foliation* are curvature surfaces. Furthermore, if the multiplicity m of κ is greater than one, then κ is constant along each of these curvature surfaces. This is not true, in general, if $m = 1$. The hypersurface M is said to be *Dupin* if along each curvature surface, the corresponding principal curvature is constant. A Dupin hypersurface is said to be *proper* if each principal curvature has constant multiplicity on M, i.e., the number of distinct principal curvatures is constant. An example of a proper Dupin surface in $\mathbb{R}^3$ is a torus of revolution. There exist many examples of Dupin hypersurfaces which are not proper, e.g., a tube M^3 in $\mathbb{R}^4$ of constant radius over a torus of revolution $T^2 \subset \mathbb{R}^3 \subset \mathbb{R}^4$ (see Section 4.2 or Pinkall [4]). The notion of Dupin can be generalized to submanifolds of higher codimension in $\mathbb{R}^n$ or even to the larger class of Legendre submanifolds. This is done in Chapter 3. Moreover, the Dupin property is easily seen to be invariant under Lie sphere transformations. This makes Lie sphere geometry a particularly effective setting for the study of Dupin submanifolds.

Non–compact proper Dupin hypersurfaces in real space–forms are plentiful. Pinkall [4] introduced four constructions for obtaining a proper Dupin hypersurface W in $\mathbb{R}^{n+m}$ from a proper Dupin hypersurface M in $\mathbb{R}^n$. These involve building tubes, cylinders, cones and surfaces of revolution from M, and they are discussed in detail in Section 4.2. Using these constructions, Pinkall was able to construct a proper Dupin hypersurface in Euclidean space with an arbitrary number of distinct principal curvatures with any given multiplicities. In general, these proper Dupin hypersurfaces cannot be extended to compact Dupin hypersurfaces without losing the property that the number of distinct principal curvatures is constant.

Compact proper Dupin submanifolds are much more rare. Obvious examples are the *isoparametric hypersurfaces* in the sphere S^n, i.e., hypersurfaces with constant principal curvatures. These were first studied by E. Cartan [2]–[5], and later by several others, especially Münzner [1], who showed that the number g of distinct principal curvatures must be 1, 2, 3, 4 or 6. Cartan classified isoparametric hypersurfaces with $g = 1$, 2 or 3 principal curvatures, but the cases $g = 4$ and $g = 6$ have not been completely classified. Ferus, Karcher and Münzner [1] produced a large class of examples with $g = 4$ using representations of Clifford algebras, many of which are not homogeneous. Their construction gives all known examples in the case $g = 4$ with two exceptions. Later Pinkall and Thorbergsson [1] gave a geometric construction of these examples. (See Sections 3.5, 3.7 and Chapter 3 of Cecil–Ryan [7] for more on isoparametric hypersurfaces.)

Thorbergsson [1] showed that the restriction $g = 1$, 2, 3, 4 or 6 on the number of distinct principal curvatures also holds for a compact proper Dupin hypersurface M embedded in $S^n \subset \mathbb{R}^{n+1}$. He first showed that M must be *taut*, i.e., every nondegenerate distance function $L_p(x) = |p - x|^2$, $p \in \mathbb{R}^{n+1}$, has the minimum number of critical points possible on M. Using tautness, he then showed that M divides S^n into two ball bundles over the first focal submanifolds on either side of M. This topological situation is all that is required for Münzner's proof of the restriction on g. Münzner's argument also produces certain restrictions on the cohomology of isoparametric hypersurfaces. These restrictions necessarily apply to compact proper Dupin hypersurfaces by Thorbergsson's result. Grove and Halperin [1] later found more topological similarities between these two classes of hypersurfaces. All of this led to the widely held conjecture that every compact proper Dupin hypersurface M embedded in S^n is equivalent by a Lie sphere transformation to an isoparametric hypersurface (see Cecil–Ryan [7, p.184]).

The conjecture is obviously true for $g = 1$, in which case M must be a sphere and is itself isoparametric. In 1978, Cecil and Ryan [2] showed that if $g = 2$, then M must be a cyclide of Dupin and is therefore Moebius equivalent to an isoparametric hypersurface. Then in 1984, Miyaoka [1] showed that the conjecture holds for $g = 3$, although it is not true that M must be Moebius equivalent to an isoparametric hypersurface. Thus, as g increases, the group needed to obtain equivalence with an isoparametric hypersurface gets progressively larger. The case $g = 4$ resisted all attempts at solution for several years and finally in 1988, counterexamples to the conjecture were discovered independently by Pinkall and Thorbergsson [1] and by Miyaoka and Ozawa [1]. The latter method also yields counterexamples with $g = 6$ principal curvatures. In both cases, a fundamental Lie invariant, the Lie curvature (Section 3.5), was used to show that the examples are not Lie equivalent to an isoparametric hypersurface. Specifically, if M is a proper Dupin hypersurface with four distinct principal curvatures, then the *Lie curvature* Ψ is the cross–ratio of these principal curvatures. Viewed in the context of projective geometry, Ψ is the cross–ratio of the four points along a projective line corresponding to the four curvature spheres of M. Hence, it is a natural Lie (projective) invariant. From the work of Münzner, it is easy to show that Ψ has the constant value $1/2$ on an isoparametric hypersurface. For the counterexamples, it was shown that $\Psi \neq 1/2$ at some points. These examples are presented in detail in Section 3.7.

Tautness was shown to be Lie invariant by Cecil and Chern [1] (see Section 3.6), and it is closely related to the Dupin condition. Pinkall [5] showed that any taut submanifold of a real space–form is Dupin. Conversely, Thorbergsson [1] proved that a compact proper Dupin hypersurface is taut. Pinkall [5] then extended this result to compact submanifolds of higher codimension for which the number of distinct principal curvatures is constant on the unit normal bundle. An open question is whether Dupin implies taut without this assumption. A key fact in Thorbergsson's proof is that in the proper Dupin case, all the curvature surfaces are spheres. In the non–proper case, the work of Ozawa [1] implies that some of the curvature surfaces are not spheres.

There is also an extensive theory of isoparametric submanifolds of arbitrary codimension in $\mathbb{R}^n$ introduced by Terng [1] (see also Harle [1]). In the case of codimension two in $\mathbb{R}^n$, this coincides with the theory of isoparametric hypersurfaces in spheres, and in codimension three, it was developed independently by Carter and West [2]–[3]. By definition, an isoparametric submanifold is necessarily Dupin. After several years of development of the theory, Thorbergsson [2] proved that all isoparametric submanifolds of codimension $m > 2$ in $\mathbb{R}^n$ are homogeneous. Thus, by the work of Palais and Terng [1], they are known to be principal orbits of isotropy representations of symmetric spaces. Heintze, Olmos and Thorbergsson [1] then gave a characterization which includes isoparametric submanifolds and their focal submanifolds as "submanifolds with constant principal curvatures." (See the survey article of Terng [5] for more on isoparametric submanifolds.)

Isoparametric submanifolds can also be characterized in terms of the L_p functions. A submanifold of $\mathbb{R}^n$ is called *totally focal* if every distance function L_p is either nondegenerate or has only degenerate critical points. Carter and West [4] proved that a compact submanifold of $\mathbb{R}^n$ is totally focal if and only if it is isoparametric. Thus, for compact submanifolds of $\mathbb{R}^n$, we have the following relationships:

$$
\begin{array}{ccc}
\text{totally focal} & \Leftrightarrow & \text{isoparametric} \\
\Downarrow & & \Downarrow \\
\text{taut} & \Rightarrow & \text{Dupin.}
\end{array}
$$

This circle of ideas would be complete if one could show that Dupin implies taut.

Chapter 4 is devoted primarily to the local classification of proper Dupin hypersurfaces in certain specific cases. These results were obtained by Lie geometric methods and have not been proven by standard Euclidean methods. In Section 4.3, we give Pinkall's [4] local classification of proper Dupin submanifolds with two distinct principal curvatures. These are known as the cyclides of Dupin. This is followed by a classification of the cyclides up to Moebius (conformal) transformation, which can be derived from the Lie classification. Finally, in Section 4.6, we present the classification of proper Dupin hypersurfaces in $\mathbb{R}^4$ with three distinct principal curvatures. This was first obtained by Pinkall [1], [3] although the treatment here is due to Cecil and Chern [2]. In the process, we develop the method of moving Lie frames which can be applied to the general study of Legendre submanifolds. This approach has been applied successfully by Niebergall [1] to obtain a partial classification of Dupin hypersurfaces in $\mathbb{R}^5$.

There are many aspects of Lie sphere geometry which are not covered in detail here. In particular, Blaschke [1] gives a more thorough treatment of the sphere geometries of Laguerre and Moebius and the "line—sphere transformation" of Lie (see Blaschke [1, §54] and Klein [1, §70]). The line—sphere transformation is discussed in a more modern setting by Fillmore [2], who also treats the relationship between Lie sphere geometry and complex line geometry. In this book, we concentrate on submanifolds of dimension greater than one in real space forms. The papers of Sasaki and Suguri [1] and Pinkall [2] treat curve theory in Lie sphere geometry. Two recent papers of Miyaoka [4]—[5] extend some of the key ideas of the Lie geometric approach to the study of contact structures and conformal structures on more general manifolds. Finally, the history and significance of Lie's early work on sphere geometry and contact transformations is discussed in the papers of Hawkins [1] and Rowe [1].

All manifolds and maps are assumed to be smooth unless explicitly stated otherwise. Notation generally follows Kobayashi and Nomizu [1] and Cecil and Ryan [7]. Theorems, equations, remarks, etc. are numbered within each chapter. If a theorem from a different chapter is cited, then the chapter is listed along with the number of the theorem.

1

Lie Sphere Geometry

In this chapter, we give Lie's construction of the space of spheres and define the important notions of oriented contact and parabolic pencils of spheres. This leads ultimately to a bijective correspondence between the manifold of contact elements on the sphere S^n and the manifold Λ^{2n-1} of projective lines on the Lie quadric.

1.1 Preliminaries

Before constructing the space of spheres, we begin with some preliminary remarks on indefinite scalar product spaces and projective geometry. Finite dimensional indefinite scalar product spaces play a crucial role in Lie sphere geometry. The fundamental result from linear algebra concerns the rank and signature of a bilinear form (see, for example, Nomizu [1, p.108], Chapter 3 of Artin [1] or O'Neill [1, pp.46–53]).

Theorem 1.1: *Suppose that* (,) *is a bilinear form on a real vector space V of dimension n. Then there exists a basis* $\{e_1,...,e_n\}$ *of V such that:*
1. $(e_i, e_j) = 0$ *for* $i \neq j$.
2. $(e_i, e_i) = 1$ *for* $1 \leq i \leq p$.
3. $(e_j, e_j) = -1$ *for* $p+1 \leq j \leq r$.
4. $(e_k, e_k) = 0$ *for* $r+1 \leq k \leq n$.

The numbers r and p are determined solely by the bilinear form; r is called the *rank*, $r-p$ is called the *index*, and the ordered pair (p, $r-p$) is called the *signature*. The theorem shows that any two spaces of the same dimension with bilinear forms of the same signature are isometrically isomorphic. A *scalar product* is a nondegenerate bilinear form, i.e., a form with rank equal to the dimension of V. For the sake of brevity, we will often refer to a scalar product as a "metric." Usually, we will be dealing with the scalar product space $\mathbb{R}^n_k$ with signature $(n–k, k)$ for $k = 0, 1$ or 2. However, at times, we will consider subspaces of $\mathbb{R}^n_k$ on which the bilinear form is degenerate. When dealing with low dimensional spaces, we will often indicate the signature with a series of plus and minus signs and zeroes where appropriate. For example, the signature of $\mathbb{R}^3_1$ may be written $(++–)$ instead of $(2, 1)$. If the bilinear form is nondegenerate, a basis with the properties listed in Theorem 1.1 is called an *orthonormal basis* for V with respect to the bilinear form.

A second useful result concerning scalar products is the following. Here $U^\perp$ denotes the orthogonal complement of a subspace U with respect to the given scalar product. (See Artin [1, p.117] or O'Neill [1, p.49].)

Theorem 1.2: *Suppose that (,) is a scalar product on a finite dimensional real vector space V and that U is a subspace of V.*

(a) *Then $U^{\perp\perp} = U$ and dim U + dim $U^\perp$ = dim V.*

(b) *The form (,) is nondegenerate on U if and only if it is nondegenerate on $U^\perp$. If the form is nondegenerate on U, then V is the direct sum of U and $U^\perp$.*

(c) *If V is the orthogonal direct sum of two subspaces U and W, then the form is nondegenerate on U and W, and $W = U^\perp$.*

Let (x, y) be the indefinite scalar product on the Lorentz space $\mathbb{R}^{n+1}_1$ defined by

$$(1.1) \qquad (x, y) = -x_1 y_1 + \ldots + x_{n+1} y_{n+1} ,$$

where $x = (x_1,...,x_{n+1})$ and $y = (y_1,...,y_{n+1})$. We will call this scalar product the *Lorentz metric*. A vector x is said to be *spacelike*, *timelike*, or *lightlike*, respectively, depending on whether (x, x) is positive, negative or zero. We will use this terminology even when we are using a metric of different signature. In Lorentz space, the set of all lightlike vectors, given by the equation,

(1.2) $$x_1^2 = x_2^2 + \ldots + x_{n+1}^2 ,$$

forms a cone of revolution, called the *light cone*. Lightlike vectors are often called *isotropic* in the literature, and the cone is called the *isotropy cone*. Timelike vectors are "inside the cone" and spacelike vectors are "outside the cone."

If x is a non–zero vector, let $x^\perp$ denote the orthogonal complement of x with respect to the Lorentz metric. If x is timelike, then the metric restricts to a positive definite form on $x^\perp$, and $x^\perp$ intersects the light cone only at the origin. If x is spacelike, then the metric has signature $(n-1, 1)$ on $x^\perp$, and $x^\perp$ intersects the cone in a cone of one less dimension. If x is lightlike, then $x^\perp$ is tangent to the cone along the line through the origin determined by x. The metric has signature $(n-1, 0)$ on this n–dimensional plane.

The true setting for Lie sphere geometry is real projective space $\mathbb{P}^n$, so we now briefly review some important concepts from projective geometry. We define an equivalence relation on $\mathbb{R}^{n+1} - \{0\}$ by setting $x \simeq y$ if $x = ty$ for some non–zero real number t. We denote the equivalence class determined by a vector x by $[x]$. $\mathbb{P}^n$ is the set of such equivalence classes, and it can naturally be identified with the space of all lines through the origin in $\mathbb{R}^{n+1}$. The rectangular coordinates $x = (x_1,\ldots,x_{n+1})$ are called *homogeneous coordinates* of the point $[x]$, and they are only determined up to a non–zero scalar multiple. The affine space $\mathbb{R}^n$ can be embedded in $\mathbb{P}^n$ as the complement of the hyperplane $(x_1 = 0)$ at infinity by the map $\varphi : \mathbb{R}^n \to \mathbb{P}^n$, $\varphi(u) = [(1, u)]$. A scalar product on $\mathbb{R}^{n+1}$, such as the Lorentz metric, determines a polar relationship between points and hyperplanes in $\mathbb{P}^n$. We will also use the notation $x^\perp$ to denote the polar hyperplane of $[x]$ in $\mathbb{P}^n$, and we will call $[x]$ the *pole* of $x^\perp$.

If x is a lightlike vector in $\mathbb{R}^{n+1}_1$, then $[x]$ can be represented by a vector of the form $(1, u)$ for $u \in \mathbb{R}^n$. Then the equation $(x, x) = 0$ for the light cone becomes $u \cdot u = 1$ (Euclidean dot product), i.e., the equation for the unit sphere in $\mathbb{R}^n$. Hence, the set of points in $\mathbb{P}^n$ determined by lightlike vectors in $\mathbb{R}^{n+1}_1$ is naturally diffeomorphic to the sphere S^{n-1}.

1.2 Moebius Geometry of Unoriented Spheres

As a first step toward Lie sphere geometry, we recall the geometry of unoriented spheres in $\mathbb{R}^n$ known as "Moebius" or "conformal" geometry. We will always assume that $n \geq 2$. In this section, we only consider spheres and planes of codimension one, and we will often omit the prefix "hyper."

We denote the Euclidean dot product of two vectors u and v in $\mathbb{R}^n$ by $u \cdot v$. We first consider stereographic projection $\sigma : \mathbb{R}^n \to S^n - \{P\}$, where S^n is the unit sphere in $\mathbb{R}^{n+1}$ given by $y \cdot y = 1$ and $P = (-1,0,...,0)$ is the south pole of S^n. (See Figure 2.1.) The well–known formula for $\sigma(u)$ is,

$$\sigma(u) = \left[\frac{1-u \cdot u}{1+u \cdot u} \, , \, \frac{2u}{1+u \cdot u} \right] .$$

We next embed $\mathbb{R}^{n+1}$ into $\mathbb{P}^{n+1}$ by the embedding φ mentioned in the previous section. Thus, we have the map $\varphi\sigma : \mathbb{R}^n \to \mathbb{P}^{n+1}$ given by,

$$(2.1) \quad \varphi\sigma\,(u) = \left[\left[1 \, , \, \frac{1-u \cdot u}{1+u \cdot u} \, , \, \frac{2u}{1+u \cdot u} \right] \right] = \left[\left[\frac{1+u \cdot u}{2} \, , \, \frac{1-u \cdot u}{2} \, , \, u \right] \right] .$$

Let $(z_1,...,z_{n+2})$ be homogeneous coordinates on $\mathbb{P}^{n+1}$ and $(\,,\,)$ the Lorentz metric on the space $\mathbb{R}_1^{n+2}$. Then $\varphi\sigma(\mathbb{R}^n)$ is just the set of points in $\mathbb{P}^{n+1}$ lying on the n–sphere Σ given by the equation $(z, z) = 0$, with the exception of the *improper point* $[(1,-1,0,...,0)]$ corresponding to the south pole P. We will refer to the points in Σ other than $[(1,-1,0,...,0)]$ as *proper points*, and will call Σ the *Moebius sphere* or *Moebius space*. At times, it is easier to simply begin with S^n rather than $\mathbb{R}^n$ and thus avoid the need for the map σ and the special point P. However, there are also advantages for beginning in $\mathbb{R}^n$.

The basic framework for the Moebius geometry of unoriented spheres is as follows. Suppose that ξ is a spacelike vector in $\mathbb{R}_1^{n+2}$. Then the polar hyperplane $\xi^\perp$ to $[\xi]$ in $\mathbb{P}^{n+1}$ intersects the sphere Σ in an $(n-1)$–sphere S^{n-1} (see Figure 2.2). S^{n-1} is the image under $\varphi\sigma$ of an $(n-1)$–sphere in $\mathbb{R}^n$, unless it contains the improper point $[(1,-1,0,...,0)]$, in which case it is the image under $\varphi\sigma$ of a hyperplane in $\mathbb{R}^n$. Hence, we have a bijective correspondence

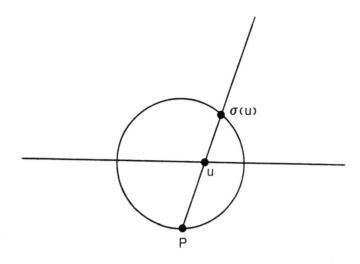

Figure 2.1 – Stereographic projection

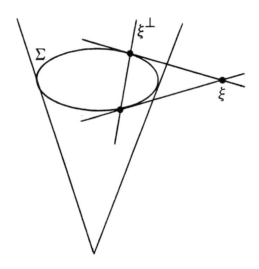

Figure 2.2 – Intersection of Σ with $\xi^{\perp}$

between the set of all spacelike points in $\mathbb{P}^{n+1}$ and the set of all hyperspheres and hyperplanes in $\mathbb{R}^n$.

It is often useful to have specific formulas for this correspondence. Consider the sphere in $\mathbb{R}^n$ with center p and radius $r > 0$ given by the equation

$$(2.2) \qquad (u - p) \cdot (u - p) = r^2 .$$

We wish to translate this into an equation involving the Lorentz metric and the corresponding polarity relationship on $\mathbb{P}^{n+1}$. A direct calculation shows that (2.2) is equivalent to the equation

$$(2.3) \qquad (\xi , \varphi\sigma (u)) = 0 ,$$

where ξ is the spacelike vector,

$$(2.4) \qquad \xi = \left[\frac{1+p \cdot p - r^2}{2}, \frac{1-p \cdot p + r^2}{2}, p \right] ,$$

and $\varphi\sigma(u)$ is given by (2.1). Thus, the point u is on the sphere (2.2) if and only if $\varphi\sigma(u)$ lies on the polar hyperplane of $[\xi]$. Note that the first two coordinates of ξ satisfy $\xi_1 + \xi_2 = 1$, and that $(\xi, \xi) = r^2$. Although ξ is only determined up to a non–zero scalar multiple, we can conclude that $\eta_1 + \eta_2$ is not zero for any $\eta \simeq \xi$.

Conversely, given a spacelike point $[z]$ with $z_1 + z_2$ non–zero, we can determine the corresponding sphere in $\mathbb{R}^n$ as follows. Let $\xi = z / (z_1 + z_2)$ so that $\xi_1 + \xi_2 = 1$. Then from (2.4), the center of the corresponding sphere is the point $p = (\xi_3,...,\xi_{n+2})$, and the radius is the square root of (ξ, ξ).

Next, suppose that η is a spacelike vector with $\eta_1 + \eta_2 = 0$. Then

$$(\eta , (1,-1,0,...,0)) = 0 .$$

Thus, the improper point $\varphi(P)$ lies on the polar hyperplane of $[\eta]$, and the point $[\eta]$ corresponds to a hyperplane in $\mathbb{R}^n$. Again we can find an explicit correspondence. Consider the hyperplane in $\mathbb{R}^n$ given by the equation

(2.5) $u \cdot N = h$, $|N| = 1.$

A direct calculation shows that (2.5) is equivalent to the equation

(2.6) $(\eta, \varphi\sigma(u)) = 0$, where $\eta = (h, -h, N)$.

Thus, the hyperplane (2.5) is represented in the polarity relationship by $[\eta]$. Conversely, let z be a spacelike point with $z_1 + z_2 = 0$. Then $(z, z) = v \cdot v$, where $v = (z_3,...,z_{n+2})$. Let $\eta = z / |v|$. Then η has the form (2.6) and $[z]$ corresponds to the hyperplane (2.5). Thus, we have explicit formulas for the bijective correspondence between the set of spacelike points in $\mathbb{P}^{n+1}$ and the set of hyperspheres and hyperplanes in $\mathbb{R}^n$.

Of course, the fundamental invariant of Moebius geometry is the angle. The study of angles in this setting is quite natural, since orthogonality between spheres and planes in $\mathbb{R}^n$ can be expressed in terms of the Lorentz metric. Let S_1 and S_2 denote the spheres in $\mathbb{R}^n$ with respective centers p_1 and p_2 and respective radii r_1 and r_2. By the Pythagorean Theorem, the two spheres intersect orthogonally (see Figure 2.3) if and only if

(2.7) $| p_1 - p_2 |^2 = r_1^2 + r_2^2 .$

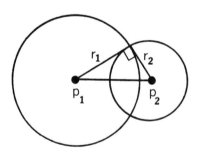

Figure 2.3 – Orthogonal spheres

If these spheres correspond by (2.4) to the projective points $[\xi_1]$ and $[\xi_2]$, respectively, then a calculation shows that (2.7) is equivalent to the condition

(2.8) $(\xi_1 , \xi_2) = 0$.

A hyperplane π in $\mathbb{R}^n$ is orthogonal to a hypersphere S precisely when π passes through the center of S. If S has center p and radius r, and π is given by the equation $u \cdot N = h$, then the condition for orthogonality is just $p \cdot N = h$. If S corresponds to $[\xi]$ as in (2.4) and π corresponds to $[\eta]$ as in (2.6), then this equation for orthogonality is equivalent to $(\xi, \eta) = 0$. Finally, if two planes π_1 and π_2 are represented by $[\eta_1]$ and $[\eta_2]$ as in (2.6), then the orthogonality condition $N_1 \cdot N_2 = 0$ is equivalent to the equation $(\eta_1, \eta_2) = 0$.

A *Moebius transformation* is a projective transformation of $\mathbb{P}^{n+1}$ which preserves the condition $(\eta, \eta) = 0$. By Theorem 1.1 of Chapter 2, a Moebius transformation also preserves the relationship $(\eta, \xi) = 0$, and it maps spacelike points to spacelike points. Thus, it preserves orthogonality (and hence angles) between spheres and planes in $\mathbb{R}^n$. In the next chapter, we will see that the group of Moebius transformations is isomorphic to $O(n+1,1)/\{\pm I\}$, where $O(n+1,1)$ is the group of orthogonal transformations of the Lorentz space $\mathbb{R}_1^{n+2}$.

Note that a Moebius transformation takes lightlike vectors to lightlike vectors and so it induces a conformal diffeomorphism of the sphere Σ onto itself. It is well known that the group of conformal diffeomorphisms of the sphere is precisely the Moebius group.

1.3 Lie Geometry of Oriented Spheres

We now turn to the construction of Lie's geometry of oriented spheres and planes in $\mathbb{R}^n$. Let W^{n+1} be the set of vectors in $\mathbb{R}_1^{n+2}$ satisfying $(\zeta, \zeta) = 1$. This is a hyperboloid of revolution of one sheet in $\mathbb{R}_1^{n+2}$. If α is a spacelike point in $\mathbb{P}^{n+1}$, then there are precisely two vectors $\pm \zeta$ in W^{n+1} with $\alpha = [\zeta]$. These two vectors can be taken to correspond to the two orientations of the oriented sphere or plane represented by α, although we have not yet given a prescription as to how to make the correspondence. To do this, we need to introduce one more coordinate. First, embed $\mathbb{R}_1^{n+2}$ into $\mathbb{P}^{n+2}$ by the embedding

$z \rightarrow [(z, 1)]$. If $\zeta \in W^{n+1}$, then

$$-\zeta_1^2 + \zeta_2^2 + \ldots + \zeta_{n+2}^2 = 1 \, ,$$

so the point $[(\zeta, 1)]$ in $\mathbb{P}^{n+2}$ lies on the quadric Q^{n+1} in $\mathbb{P}^{n+2}$ given in homogeneous coordinates by the equation

(3.1) $< x, x > = - x_1^2 + x_2^2 + \ldots + x_{n+2}^2 - x_{n+3}^2 = 0 \, .$

Q^{n+1} is called the *Lie quadric*, and the scalar product determined by the quadratic form in (3.1) is called the *Lie metric* or *Lie scalar product*. We will let $\{e_1,\ldots,e_{n+3}\}$ denote the standard orthonormal basis for the scalar product space $\mathbb{R}_2^{n+3}$ with metric $< , >$. Here e_1 and e_{n+3} are timelike and the rest are spacelike.

We shall now see how the points on Q^{n+1} correspond to the set of oriented spheres, oriented planes and point spheres of $\mathbb{R}^n \cup \{\infty\}$. Suppose that x is any point on the quadric with homogeneous coordinate $x_{n+3} \neq 0$. Then x can be represented by a vector of the form $(\zeta, 1)$, where the Lorentz scalar product $(\zeta, \zeta) = 1$. Suppose first that $\zeta_1 + \zeta_2 \neq 0$. Then, in Moebius geometry, $[\zeta]$ represents a sphere in $\mathbb{R}^n$. If as in (2.4), we represent $[\zeta]$ by a vector

$$\xi = \left[\frac{1 + p \cdot p - r^2}{2} , \frac{1 - p \cdot p + r^2}{2} , p \right] \, ,$$

then $(\xi, \xi) = r^2$. Thus, ζ must be one of the vectors $\pm \xi / r$. In $\mathbb{P}^{n+2}$,

$$[(\zeta, 1)] = [(\pm \xi / r, 1)] = [(\xi, \pm r)] \, .$$

We can interpret the last coordinate as a signed radius of the sphere with center p and unsigned radius $r > 0$. In order to be able to interpret this geometrically, we adopt the convention that a positive signed radius corresponds to the orientation of the sphere represented by the inward field of unit normals, and a negative signed radius corresponds to the orientation given by the outward field of unit normals. Hence, the two orientations of the sphere in $\mathbb{R}^n$ with center p and unsigned radius $r > 0$ are represented by the two projective points,

(3.2)
$$\left[\left[\frac{1+p \cdot p - r^2}{2}, \frac{1-p \cdot p + r^2}{2}, \pm r\right]\right],$$

in Q^{n+1}. Next, if $\zeta_1 + \zeta_2 = 0$, then $[\zeta]$ represents a hyperplane in $\mathbb{R}^n$, as in (2.6). For $\zeta = (h, -h, N)$, with $|N| = 1$, we have $(\zeta, \zeta) = 1$. Then the two projective points on Q^{n+1} induced by ζ and $-\zeta$ are,

(3.3)
$$[(h, -h, N, \pm 1)] .$$

These represent the two orientations of the plane with equation $u \cdot N = h$. We make the convention that $[(h, -h, N, 1)]$ corresponds to the orientation of this plane given by the field of unit normals N, while the orientation given by $- N$ corresponds to the point $[(h, -h, N, -1)] = [(-h, h, - N, 1)]$.

Thus far, we have determined a bijective correspondence between the set of points x in Q^{n+1} with $x_{n+3} \neq 0$ and the set of all oriented spheres and planes in $\mathbb{R}^n$. Suppose now that $x_{n+3} = 0$, i.e., consider a point $[(z, 0)]$, for z in $\mathbb{R}_1^{n+2}$. Then, $< x, x > = (z, z) = 0$, and $[z]$ in $\mathbb{P}^{n+1}$ is simply a point of the Moebius sphere Σ. Thus, we have the following bijective correspondence:

Euclidean	Lie
Points: $u \in \mathbb{R}^n$	$\left[\left[\dfrac{1+u \cdot u}{2}, \dfrac{1-u \cdot u}{2}, u, 0\right]\right],$
∞	$[(1, -1, 0, 0)],$
Spheres: Center p, signed radius r	$\left[\left[\dfrac{1+p \cdot p - r^2}{2}, \dfrac{1-p \cdot p + r^2}{2}, p, r\right]\right],$
Planes: $u \cdot N = h$, unit normal N	$[(h, -h, N, 1)] .$

(3.4)

In Lie sphere geometry, points are considered to be spheres of radius zero, or point spheres. From now on, we will use the term "Lie sphere" or simply "sphere" to denote an oriented sphere, oriented plane or a point sphere in $\mathbb{R}^n \cup \{\infty\}$. We will refer to the coordinates on the right side of the chart above as the *Lie coordinates* of the corresponding point, sphere or plane. In the case of $\mathbb{R}^2$ and $\mathbb{R}^3$, respectively, these coordinates were classically called *pentaspherical* and *hexaspherical* coordinates (see Blaschke [1]). At times, it is useful to have formulas to convert the Lie coordinates back into Cartesian equations for the corresponding object in $\mathbb{R}^n$. Suppose first that $[x]$ is a point on the quadric with $x_1 + x_2 \neq 0$. Then, $x = \rho y$, for some $\rho \neq 0$, where y is one of the standard forms in (3.4). In (3.4), we see that $y_1 + y_2 = 1$, for all proper points and all spheres. Hence, if we divide x by $x_1 + x_2$, the new vector will be in standard form, and we can read off the Cartesian data from (3.4). In particular, if $x_{n+3} = 0$, then $[x]$ represents the point $u = (u_3,...,u_{n+2})$ where

$$(3.5) \qquad u_i = x_i / (x_1 + x_2) , \quad 3 \leq i \leq n+2 .$$

If $x_{n+3} \neq 0$, then $[x]$ represents the sphere with center $p = (p_3,...,p_{n+2})$ and signed radius r given by

$$(3.6) \qquad p_i = x_i / (x_1 + x_2) , \quad 3 \leq i \leq n+2 ; \qquad r = x_{n+3} / (x_1 + x_2) .$$

Finally, suppose that $x_1 + x_2 = 0$. If $x_{n+3} = 0$, then the condition $< x, x > = 0$ forces x_i to be zero for $3 \leq i \leq n+2$. Thus, $[x] = [(1,-1,0,...,0)]$, the improper point. If $x_{n+3} \neq 0$, we divide x by x_{n+3} to make the last coordinate 1. Then, if we set $N = (N_3,...,N_{n+2})$ and h according to

$$(3.7) \qquad N_i = x_i / x_{n+3} , \quad 3 \leq i \leq n+2 ; \qquad h = x_1 / x_{n+3} ,$$

the conditions $< x, x > = 0$ and $x_1 + x_2 = 0$, force N to have unit length. Thus, $[x]$ corresponds to the hyperplane $u \cdot N = h$, with unit normal N and h as in (3.7).

1.4 Geometry of Hyperspheres in S^n and H^n

In some ways, it is simpler to use S^n rather than $\mathbb{R}^n$ as the base space for the study of Moebius or Lie sphere geometry. This avoids the use of stereographic projection and the need to refer to an improper point or to distinguish between spheres and planes. Furthermore, the correspondence given in equation (3.4) can be reduced to a single formula (4.4) below.

As in Section 1.2, we consider S^n to be the unit sphere in $\mathbb{R}^{n+1}$, and then embed $\mathbb{R}^{n+1}$ into $\mathbb{P}^{n+1}$ by the canonical embedding φ. Then $\varphi(S^n)$ is the Moebius sphere Σ, given by the equation $(z, z) = 0$ in homogeneous coordinates. First, we find the Moebius equation for the unoriented hypersphere in S^n with center $p \in S^n$ and spherical radius ρ, $0 < \rho < \pi$. This hypersphere is the intersection of S^n with the hyperplane in $\mathbb{R}^{n+1}$ given by the equation

(4.1) $$p \cdot y = \cos \rho, \quad 0 < \rho < \pi.$$

Let $[z] = \varphi(y) = [(1, y)]$. Then

$$p \cdot y = \frac{-(z, (0, p))}{(z, e_1)}.$$

Thus, the equation (4.1) can be rewritten as

(4.2) $$(z, (\cos \rho, p)) = 0.$$

Therefore, a point $y \in S^n$ is on the hyperplane determined by (4.1) if and only if $\varphi(y)$ lies on the polar hyperplane in $\mathbb{P}^{n+1}$ of the point

(4.3) $$[\xi] = [(\cos \rho, p)].$$

To obtain the two oriented spheres determined by (4.1) note that

$$(\xi, \xi) = -\cos^2 \rho + 1 = \sin^2 \rho.$$

Noting that $\sin \rho \neq 0$, we let $\zeta = \pm \xi / \sin \rho$. Then the point $[(\zeta, 1)]$ is on Q^{n+1}, and

$$[(\zeta, 1)] = [(\xi, \pm \sin \rho)] = [(\cos \rho, p, \pm \sin \rho)] .$$

We can incorporate the sign of the last coordinate into the radius and thereby arrange that the oriented sphere with signed radius $\rho \neq 0$, $-\pi < \rho < \pi$, and center p corresponds to the point in Q^{n+1},

(4.4) $[(\cos \rho, p, \sin \rho)] .$

The formula still makes sense if the radius $\rho = 0$, in which case it yields the point sphere $[(1, p, 0)]$. This one formula (4.4) plays the role of all the formulas given in (3.4) for the Euclidean case.

As in the Euclidean case, the orientation of a sphere S in S^n is determined by a choice of unit normal field to S in S^n. Geometrically, we take positive radius in (4.4) to correspond to the field of unit normals which are tangent vectors to geodesics in S^n from p to $-p$. Each oriented sphere can be considered in two ways, with center p and signed radius ρ, $-\pi < \rho < \pi$, or with center $-p$ and the appropriate signed radius $\rho \pm \pi$.

Given a point $[x] \in Q^{n+1}$, we now determine the corresponding hypersphere in S^n. Multiplying by -1, if necessary, we may assume that the first homogeneous coordinate x_1 of x satisfies $x_1 \geq 0$. If $x_1 > 0$, then we see from (4.4) that the center p and signed radius ρ, $-\pi/2 < \rho < \pi/2$, satisfy

(4.5) $\tan \rho = x_{n+3} / x_1, \quad p = (x_2,...,x_{n+2}) / (x_1^2 + x_{n+3}^2)^{1/2} .$

If $x_1 = 0$, then $x_{n+3} \neq 0$, so we can divide by x_{n+3} to obtain a point of the form $(0, p, 1)$. This corresponds to the oriented hypersphere with center p and signed radius $\pi/2$, which is a great sphere in S^n.

To treat oriented hyperspheres in hyperbolic space H^n, we let $\mathbb{R}_1^{n+1}$ be the Lorentz subspace of $\mathbb{R}_1^{n+2}$, spanned by the orthonormal basis $\{e_1, e_3,...,e_{n+2}\}$. Then, H^n is the hypersurface

$$\{ \, y \in \mathbb{R}_1^{n+1} \mid (y, y) = -1 \, , \; y_1 \geq 1 \, \} \, ,$$

on which the restriction of the Lorentz metric $(\,,\,)$ is a positive definite metric of constant sectional curvature -1 (see Kobayashi–Nomizu [1, Vol. II, pp. 268–271] for more detail). The distance between two points p and q in H^n is given by $d(\, p, q \,) = \cosh^{-1}(- (\, p, q \,))$. Thus, the equation for the unoriented sphere in H^n with center p and radius ρ is

(4.6) $(\, p, y \,) = - \cosh \rho \, .$

As before with S^n, we first embed $\mathbb{R}_1^{n+1}$ into $\mathbb{P}^{n+1}$ by the map $\psi(y) = [(\, y + e_2 \,)]$. Let $p \in H^n$ and let $z = y + e_2$ for $y \in H^n$. Then we have

$$(\, p, y \,) = (\, z, p \,) \, / \, (\, z, e_2 \,) \, .$$

Hence, equation (4.6) is equivalent to the condition that $[z] = [(\, y + e_2 \,)]$ lies on the polar hyperplane in $\mathbb{P}^{n+1}$ to $[\xi] = [(\, p + \cosh \rho \; e_2)]$. Following exactly the same procedure as in the spherical case, we find that the oriented hypersphere in H^n with center p and signed radius ρ corresponds to the point in Q^{n+1} given by

(4.7) $[(\, p + \cosh \rho \; e_2 + \sinh \rho \; e_{n+3} \,)] \, .$

There is also a stereographic projection τ with pole $- e_1$ from H^n onto the unit disk D^n in $\mathbb{R}^n = \mathrm{Span} \, \{e_3,...,e_{n+2}\}$ given by

(4.8) $\tau(\, y_1, y_3,...,y_{n+2} \,) = (\, y_3,...,y_{n+2} \,) / (\, y_1 + 1 \,).$

The metric g induced on D^n in order to make τ an isometry is the usual Poincaré metric.

In Section 2.5, we will see that from the point of view of Klein's Erlangen Program, all three of these geometries, Euclidean, spherical and hyperbolic, are subgeometries of Lie sphere geometry.

1.5 Oriented Contact and Parabolic Pencils of Spheres

In Moebius geometry, the principal geometric quantity is the angle. In Lie sphere geometry, the corresponding central concept is that of oriented contact of spheres. Two oriented spheres S_1 and S_2 in $\mathbb{R}^n$ are in *oriented contact* if they are tangent and if they have the same orientation at the point of contact. (See Figure 5.1 for the two possibilities.) If p_1 and p_2 are the respective centers of S_1 and S_2, and r_1 and r_2 are their respective signed radii, then the analytic condition for oriented contact is

$$(5.1) \qquad\qquad | p_1 - p_2 | = | r_1 - r_2 | \, .$$

An oriented sphere S with center p and signed radius r is in oriented contact with an oriented hyperplane π with unit normal N and equation $u \cdot N = h$ if π is tangent to S and their orientations agree at the point of contact. Analytically, this is just the equation

$$(5.2) \qquad\qquad p \cdot N = r + h \, .$$

Two oriented planes π_1 and π_2 are in oriented contact if their unit normals N_1 and N_2 are the same. Two such planes can be thought of as two oriented spheres in oriented contact at the improper point.

A proper point u in $\mathbb{R}^n$ is in oriented contact with an oriented sphere or plane if it lies on the sphere or plane. Analytically, this means that u satisfies the equation defining the sphere or plane. Finally, the improper point is in oriented contact with each plane, since it lies on each plane.

Suppose that S_1 and S_2 are two Lie spheres which are represented in the standard form (3.4) by $[k_1]$ and $[k_2]$. One can check directly that in all cases, the analytic condition for oriented contact is equivalent to the equation

$$(5.3) \qquad\qquad < k_1 , k_2 > = 0 \, .$$

Next, we do some linear algebra to establish the important fact that the Lie quadric contains projective lines but no linear subspaces of higher

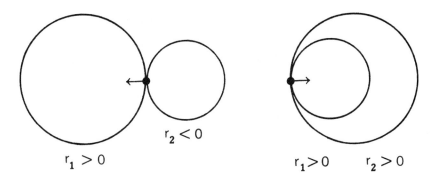

Figure 5.1 — Oriented contact of spheres

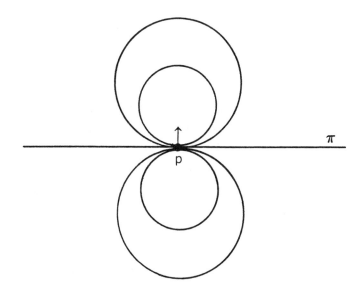

Figure 5.2 — Parabolic pencil of spheres

dimension. We then show that the set of oriented spheres in $\mathbb{R}^n$ corresponding to the points on a line on Q^{n+1} forms a so–called *parabolic pencil* of spheres (see Figure 5.2). We also show that each parabolic pencil contains exactly one point sphere. Furthermore, if this point sphere is a proper point p in $\mathbb{R}^n$, then the pencil contains exactly one hyperplane π. The pencil consists of all oriented hyperspheres in oriented contact with π at the point p.

The fundamental result needed from linear algebra is the following. Note, a subspace of a scalar product space is called *lightlike* if it consists of only lightlike vectors.

Theorem 5.1: *Let (,) be a scalar product of signature (n–k, k) on a real vector space V. Then the maximal dimension of a lightlike subspace of V is the minimum of the numbers k and n–k.*

Proof: First, note that the theorem holds for scalar products having signature $(n–k, k)$ if and only if it holds for scalar products of signature $(k, n–k)$, since changing the signs of the quantities (e_i, e_i) for an orthonormal basis does not change the set of lightlike vectors.

Thus, we now assume that $k \leq n–k$ and do the proof by induction on the index k. The theorem is clearly true for scalar products of index 0, since the only lightlike vector is 0 itself. Assume now that the theorem holds for all spaces with a scalar product of index $k–1$, and let V be a scalar product space of index $k \geq 1$. Let W be a lightlike subspace of V of maximal dimension, and let v be a timelike vector in V. Then the scalar product restricts to a scalar product of index $k–1$ on the hyperplane $U = v^{\perp}$, and $W \cap U$ is a lightlike subspace of W. By the induction hypothesis, dim $W \cap U \leq k–1$ and therefore, dim $W \leq k$, as desired. On the other hand, it is easy to exhibit a lightlike subspace of V of dimension k. Let $\{e_1,...,e_n\}$ be an orthonormal basis for V with $e_1,...,e_k$ timelike and the rest spacelike. For $1 \leq i \leq k$, let $v_i = e_i + e_{k+i}$. Then the span of $v_1,...,v_k$ is a lightlike subspace of dimension k.   □

Corollary 5.2: *The Lie quadric contains projective lines but no linear subspaces of higher dimension.*

Proof: This follows immediately from Theorem 5.1, since a linear subspace of $\mathbb{P}^{n+2}$ of dimension $k-1$ which lies on the quadric corresponds to a lightlike subspace of dimension k in $\mathbb{R}_2^{n+3}$.   □

Theorem 5.1 also implies the following result concerning the orthogonal complement of a line on the quadric. This was pointed out by Pinkall [1, p.24] .

Corollary 5.3: *Let ℓ be a line in $\mathbb{P}^{n+2}$ which lies on Q^{n+1}.*
(a)   *If $[x] \in \ell^{\perp}$ and $[x]$ is lightlike, then $[x] \in \ell$.*
(b)   *If $[x] \in \ell^{\perp}$ and $[x]$ is not on ℓ, then $[x]$ is spacelike.*

Proof: (a) Suppose that $[x]$ is a lightlike point in $\ell^{\perp}$ but not on ℓ. Then the 2–dimensional linear lightlike subspace spanned by $[x]$ and ℓ lies on the quadric, contradicting Corollary 5.2.
(b) Suppose that $[x]$ is in $\ell^{\perp}$ but not on ℓ. From (a) we know that $[x]$ is either spacelike or timelike. Suppose that $[x]$ is timelike. Then the Lie metric $< , >$ has signature $(n+1,1)$ on the vector space $x^{\perp}$, and $x^{\perp}$ contains the 2–dimensional lightlike vector space which projects to ℓ. This contradicts Theorem 5.1.   □

The next result establishes the relationship between the points on a line in Q^{n+1} and the corresponding pencil of spheres in $\mathbb{R}^{n}$.

Theorem 5.4: (a) *The line in $\mathbb{P}^{n}$ determined by two points $[k_1]$ and $[k_2]$ of Q^{n+1} lies on Q^{n+1} if and only if the spheres corresponding to $[k_1]$ and $[k_2]$ are in oriented contact, i.e., $< k_1, k_2 > = 0$.*
(b) *If the line $[k_1, k_2]$ lies on Q^{n+1}, then the parabolic pencil of spheres in $\mathbb{R}^{n}$ corresponding to points on $[k_1, k_2]$ is precisely the set of all spheres in oriented contact with both $[k_1]$ and $[k_2]$.*

Proof: (a) The line $[k_1, k_2]$ consists of the points of the form $[\alpha k_1 + \beta k_2]$, where α and β are any two real numbers, at least one of which is not 0. Since $[k_1]$ and $[k_2]$ are on Q^{n+1}, we have

$$< \alpha k_1 + \beta k_2 , \alpha k_1 + \beta k_2 > = 2 \alpha \beta < k_1 , k_2 > .$$

Thus, the line is contained in the quadric if and only if $< k_1, k_2 > = 0$.

(b) Let $[\alpha k_1 + \beta k_2]$ be any point on the line. Since $< k_1 , k_2 > = 0$ by (a), we easily compute that $[\alpha k_1 + \beta k_2]$ is orthogonal to both $[k_1]$ and $[k_2]$. Hence, the corresponding sphere is in oriented contact with the spheres corresponding to $[k_1]$ and $[k_2]$. Conversely, suppose that the sphere corresponding to a point $[k]$ on the quadric is in oriented contact with the spheres corresponding to $[k_1]$ and $[k_2]$. Then $[k]$ is orthogonal to every point on the line $[k_1, k_2]$, and so $[k]$ is on the line by Corollary 5.3 (a).

$\square$

As we have noted in the proofs of the previous results, given any timelike point $[z]$ in $\mathbb{P}^{n+2}$, the scalar product $< , >$ has signature $(n+1,1)$ on $z^{\perp}$. Hence, $z^{\perp}$ intersects Q^{n+1} in a Moebius space. We now show that any line on the quadric intersects such a Moebius space in exactly one point.

Corollary 5.5: *Let* $[z]$ *be a timelike point in* $\mathbb{P}^{n+2}$ *and* ℓ *a line which lies on* Q^{n+1}. *Then* ℓ *intersects* $z^{\perp}$ *in exactly one point.*

Proof: Any line in projective space intersects a hyperplane in at least one point. We simply must show that ℓ is not contained in $z^{\perp}$. But this follows from Theorem 5.1, since $< , >$ has signature $(n+1, 1)$ on $z^{\perp}$, and therefore $z^{\perp}$ cannot contain the 2–dimensional lightlike vector space which projects to ℓ. $\square$

As a consequence, we obtain the following corollary.

Corollary 5.6: *Every parabolic pencil contains exactly one point sphere. Furthermore, if the point sphere is a proper point, then the pencil contains exactly one plane.*

Proof: The point spheres are precisely the points of intersection of Q^{n+1} with $e_{n+3}^{\perp}$. Thus, each parabolic pencil contains exactly one point sphere by Corollary 5.5. The hyperplanes correspond to the points in the intersection of Q^{n+1} with $(e_1 - e_2)^{\perp}$. The line ℓ on the quadric corresponding to the given parabolic pencil intersects this hyperplane in exactly one point, unless ℓ is contained in the hyperplane. But ℓ is contained in $(e_1 - e_2)^{\perp}$ if and only if the improper point $[e_1 - e_2]$ is in $\ell^{\perp}$. By Corollary 5.3 (a), this implies that the point $[e_1 - e_2]$ is on ℓ. Hence, if the point sphere of the pencil is not the improper point, then the pencil contains exactly one hyperplane.   □

By Corollary 5.6 and Theorem 5.4, we see that if the point sphere in a parabolic pencil is a proper point $p \in \mathbb{R}^n$, then the pencil consists precisely of all spheres in oriented contact with a certain oriented plane π at p. Thus, one can identify the parabolic pencil with the point (p, N) in the unit tangent bundle to $\mathbb{R}^n$, where N is the unit normal to the plane π. If the point sphere of the pencil is the improper point, then the pencil must consist entirely of planes. Since these planes are all in oriented contact, they all have the same normal N. Thus, the pencil can be identified with the point (∞, N) in the unit tangent bundle to $\mathbb{R}^n \cup \{\infty\} = S^n$.

It is also useful to have this correspondence between parabolic pencils and elements of the unit tangent bundle $T_1 S^n$ expressed in terms of the spherical metric on S^n. Suppose that ℓ is a line on the quadric. From Corollary 5.5 and equation (4.4), we see that ℓ intersects both $e_1^{\perp}$ and $e_{n+3}^{\perp}$ in exactly one point. So the corresponding parabolic pencil contains exactly one point sphere and one great sphere, represented respectively by the points,

$$[k_1] = [(1, p, 0)] , \quad [k_2] = [(0, \xi, 1)] .$$

The fact that $< k_1, k_2 > = 0$ is equivalent to the condition $p \cdot \xi = 0$, i.e., ξ is tangent to S^n at p. Hence the parabolic pencil of spheres corresponding to ℓ can be identified with the point (p, ξ) in $T_1 S^n$. The points on the line ℓ can be parametrized as

$$[K_t] = [\cos t \, k_1 + \sin t \, k_2] = [(\cos t , \cos t \, p + \sin t \, \xi , \sin t)] .$$

From (4.4), we see that $[K_t]$ corresponds to the sphere in S^n with center

(5.4) $$p_t = \cos t \, p + \sin t \, \xi ,$$

and signed radius t. These are precisely the spheres through p in oriented contact with the great sphere corresponding to $[k_2]$. Their centers lie along the geodesic in S^n with initial point p and initial velocity vector ξ .

2

Lie Sphere Transformations

In this chapter, we study the sphere geometries of Lie, Moebius and Laguerre from the point of view of Klein's Erlangen Program. In each case, we determine the group of transformations which preserve the fundamental geometric properties of the space. All of these groups are quotient groups or subgroups of some orthogonal group determined by an indefinite scalar product on a real vector space. As a result, the theorem of Cartan and Dieudonné, proven in Section 2.2, implies that each of these groups is generated by inversions. In Section 2.3, we give a geometric description of Moebius inversions. This is followed by a treatment of Laguerre geometry in Section 2.4. Finally, in Section 2.5, we show that the Lie sphere group is generated by the union of the groups of Moebius and Laguerre. There we also describe the place of Euclidean, spherical and hyperbolic metric geometries within the context of these more general geometries.

2.1 The Fundamental Theorem

A *Lie sphere transformation* is a projective transformation of P^{n+2} which takes Q^{n+1} to itself. In terms of the geometry of $\mathbb{R}^n$, a Lie sphere transformation maps oriented spheres to oriented spheres. (Here the term "sphere" includes planes and point spheres.) Furthermore, since it is projective, a Lie sphere transformation maps lines on Q^{n+1} to lines on Q^{n+1}. Thus, it preserves oriented contact of spheres in $\mathbb{R}^n$. We will first show that the group G of Lie sphere transformations is isomorphic to $O(n+1,2)/\{\pm I\}$, where $O(n+1,2)$ is the group of orthogonal transformations of $\mathbb{R}_2^{n+3}$. We will then give Pinkall's [4] proof of the so-called "Fundamental Theorem of Lie sphere geometry" which states that any line preserving diffeomorphism of Q^{n+1} is the restriction to Q^{n+1} of a projective transformation, that is, a transformation of the space of spheres which preserves oriented contact must be a Lie transformation.

Recall that a linear transformation A in $GL(n+1)$ induces a projective transformation $P(A)$ on $\mathbb{P}^n$ defined by $P(A)[x] = [Ax]$. The map P is a homomorphism of $GL(n+1)$ onto the group $PGL(n)$ of projective transformations of $\mathbb{P}^n$. It is well known (see, for example, Samuel [1, p.6]) that the kernel of P is the group of all non–zero scalar multiples of the identity transformation.

The fact that the group G is isomorphic to $O(n+1,2)/\{\pm I\}$ follows immediately from the following theorem. Here we let $<,>$ denote the scalar product on $\mathbb{R}^n_k$.

Theorem 1.1: *Let A be a nonsingular linear transformation on the indefinite scalar product space $\mathbb{R}^n_k$, $1 \leq k \leq n-1$, such that A takes lightlike vectors to lightlike vectors.*

(a)   *Then there is a non-zero constant λ such that $<Av, Aw> = \lambda <v, w>$ for all v, w in $\mathbb{R}^n_k$.*

(b)   *Furthermore, if $k \neq n-k$, then $\lambda > 0$.*

Proof: Since $k \geq 1$ and $n-k \geq 1$, there exist both timelike and spacelike vectors. Suppose that v is a unit timelike vector and w is a unit spacelike vector such that $<v, w> = 0$. Then $v+w$ and $v-w$ are both lightlike. By the hypothesis of the theorem, $A(v+w)$ and $A(v-w)$ are both lightlike. Thus, we have

$$0 = <A(v+w), A(v+w)> = <Av, Av> + 2<Av, Aw> + <Aw, Aw>,$$
(1.1)
$$0 = <A(v-w), A(v-w)> = <Av, Av> - 2<Av, Aw> + <Aw, Aw>.$$

If we subtract the second from the first, we get $<Av, Aw> = 0$. Substitution of this into either equation in (1.1) yields

(1.2) $-<Av, Av> = <Aw, Aw> = \lambda,$

for some real number λ. Now, suppose that $v_1,...,v_k,w_1,...,w_{n-k}$ is an orthonormal basis for $\mathbb{R}^n_k$ with $v_1,...,v_k$ timelike and $w_1,...,w_{n-k}$ spacelike. We have already shown that $< Av_i , Aw_j > = 0$ for all i and j. From (1.2), we get that $- < Av_i , Av_i > = < Aw_j , Aw_j > = \lambda$ for all i, j, since we can first hold v constant in (1.2) and vary w, then hold w constant and vary v. It remains to be shown that $< Av_i , Av_j > = 0$ and $< Aw_i , Aw_j > = 0$ for $i \neq j$. Consider the vector $w = (w_i + w_j) /\sqrt{2}$. Then w is a unit spacelike vector orthogonal to v_1. Hence, we have $< Aw, Aw > = \lambda$, i.e.,

$$(1.3) \qquad 2\lambda = < A(w_i + w_j), A(w_i + w_j) >$$
$$= < Aw_i , Aw_i > + 2 < Aw_i , Aw_j > + < Aw_j , Aw_j >.$$

Since $< Aw_i , Aw_i > = < Aw_j , Aw_j > = \lambda$, we have $< Aw_i , Aw_j > = 0$ for $i \neq j$. A similar proof shows that $< Av_i , Av_j > = 0$ for $i \neq j$. Therefore, the equation $< Ax, Ay > = \lambda < x, y >$ holds on an orthonormal basis, so it holds for all vectors.

To prove (b), note that $< , >$ has signature $(k, n-k)$; so if $k \neq n-k$, then the Av_i must be timelike and the Aw_i spacelike, i.e., $\lambda > 0$.   □

Remark 1.2: In the case $k = n-k$, conclusion (b) does not necessarily hold. For example the linear map T defined by $Tv_i = w_i$, $Tw_i = v_i$, for $1 \leq i \leq k$, preserves lightlike vectors, but the corresponding $\lambda = -1$.

From Theorem 1.1, we immediately obtain the following corollary.

Corollary 1.3: (a) *The group G of Lie sphere transformations is isomorphic to* $O(n+1,2)/\{\pm I\}$.
(b) *The group H of Moebius transformations is isomorphic to* $O(n+1,1)/\{\pm I\}$.

Proof: (a) Suppose $\alpha = P(A)$ is a Lie sphere transformation. By Theorem 1.1, we have $< Av, Aw > = \lambda < v, w >$ for all v, w in $\mathbb{R}^{n+3}_2$, where λ is a positive constant. Set B equal to $A/\sqrt{\lambda}$. Then B is in $O(n+1,2)$ and $\alpha = P(B)$. Thus, every Lie sphere transformation can be represented by an orthogonal transformation. Conversely, if $B \in O(n+1,2)$, then $P(B)$ is clearly a Lie sphere transformation. Now, let $\Psi : O(n+1,2) \to G$ be the restriction of the homomorphism P to $O(n+1,2)$. Then Ψ is surjective with kernel equal to the intersection of the kernel of P with $O(n+1,2)$, i.e., kernel $\Psi = \{\pm I\}$.

(b) This follows from Theorem 1.1 in the same manner as (a) with the Lorentz metric being used instead of the Lie metric.   □

Remark 1.4: *On Moebius transformations in Lie sphere geometry.*

A Moebius transformation is a transformation on the space of unoriented spheres, i.e., the space of projective classes of spacelike vectors in $\mathbb{R}^{n+2}_1$. Hence, each Moebius transformation naturally induces two Lie transformations on the space Q^{n+1} of oriented spheres. Specifically, if A is in $O(n+1,1)$, then we can extend A to a transformation B in $O(n+1,2)$ by setting $B = A$ on $\mathbb{R}^{n+2}_1$ and $B(e_{n+3}) = e_{n+3}$. In terms of matrix representation with respect to the standard orthonormal basis, B has the form

$$(1.4) \qquad\qquad B = \begin{bmatrix} A & 0 \\ 0 & 1 \end{bmatrix}.$$

Note that while A and $-A$ induce the same Moebius transformation, the Lie transformation $P(B)$ is not the same as the Lie transformation $P(C)$ induced by the matrix

$$C = \begin{bmatrix} -A & 0 \\ 0 & 1 \end{bmatrix} \cong \begin{bmatrix} A & 0 \\ 0 & -1 \end{bmatrix},$$

where $\cong$ denotes equivalence as projective transformations. Hence, the Moebius transformation $P(A) = P(-A)$ induces two Lie transformations, $P(B)$ and $P(C)$. Finally, note that $P(B) = \Gamma P(C)$, where Γ is the Lie transformation represented in matrix form by

$$\Gamma = \begin{bmatrix} I & 0 \\ 0 & -1 \end{bmatrix} \cong \begin{bmatrix} -I & 0 \\ 0 & 1 \end{bmatrix}.$$

From equation (3.4) of Chapter 1, we see that Γ has the effect of changing the orientation of every oriented sphere and plane. We will call Γ the *change of orientation transformation*, although the German word "Richtungswechsel" is certainly more economical. Hence, the two Lie transformations induced by the Moebius transformation $P(A)$ differ by this change of orientation factor. Thus, the group of Lie transformations induced from Moebius transformations is isomorphic to $O(n+1,1)$ and is a double covering of the Moebius group H. This group consists of those Lie transformations which map $[e_{n+3}]$ to itself. Since such a transformation must also take $e_{n+3}^{\perp}$ to itself, this is precisely the group of Lie transformations which take point spheres to point spheres. When working in the context of Lie geometry, we will often refer to these transformations as "Moebius transformations."

We now turn to the proof of the Fundamental Theorem. Lie [1, p.186] proved this result in his thesis for $n = 2$. (See also Lie–Scheffers [1, p.437] or Blaschke [1, p.211].) The following proof is due to Pinkall [4, p.431].

Theorem 1.5: *Every line preserving diffeomorphism of Q^{n+1} is the restriction to Q^{n+1} of a Lie sphere transformation.*

The key observation in this proof is that a line–preserving diffeomorphism of Q^{n+1} corresponds to a conformal transformation of Q^{n+1} endowed with its natural pseudo–Riemannian metric of signature $(n, 1)$. Theorem 1.5 then follows from the fact that such a conformal transformation must be the restriction to Q^{n+1} of a projective transformation of $\mathbb{P}^{n+2}$ induced by an orthogonal transformation of $\mathbb{R}_2^{n+3}$. This is a generalization of the fact that a conformal transformation of S^n must be a Moebius transformation, since S^n is just the quadric obtained by projection of the light cone in $\mathbb{R}_1^{n+2}$ (see Section 1.2). In fact, given any scalar product space $\mathbb{R}_{k+1}^{m+2}$, the projective quadric Q_k^m obtained by projecting the light cone in $\mathbb{R}_{k+1}^{m+2}$ has a natural pseudo–Riemannian metric of signature $(m-k, k)$. With this metric Q_k^m is a conformally flat pseudo–Riemannian symmetric space, called the standard *projective quadric* of signature $(m-k, k)$. (In this notation, our Q^{n+1} is Q_1^{n+1} and S^n is Q_0^n.) Cahen and Kerbrat [1, pp.327–331] give a proof that the group of conformal transformations of Q_k^m is isomorphic to $O(m-k+1, k+1)/\{\pm I\}$ which works for all signatures at once. Here, we will construct the standard metric for our quadric Q^{n+1} and demonstrate that a line–preserving diffeomorphism of Q^{n+1} determines a conformal transformal of Q^{n+1} with respect to this metric. Let V^{n+2} be the light cone in $\mathbb{R}_2^{n+3}$, and let

$$M^{n+1} = \{ x \in \mathbb{R}_2^{n+3} \mid x_1^2 + x_{n+3}^2 = 1, \quad x_2^2 + ... + x_{n+2}^2 = 1 \} .$$

M^{n+1} is the intersection of V^{n+2} with the hypersphere of radius $\sqrt{2}$ in the Euclidean metric on $\mathbb{R}^{n+3}$. It is clearly diffeomorphic to $S^1 \times S^n$, where S^1 is the circle $x_1^2 + x_{n+3}^2 = 1$ in the timelike plane spanned by e_1 and e_{n+3}, and S^n is the sphere

$$x_2^2 + ... + x_{n+2}^2 = 1 ,$$

in the Euclidean space $\mathbb{R}^{n+1}$ spanned by $e_2,...,e_{n+2}$. The manifold M^{n+1} is a double covering of Q^{n+1}, and the fiber containing a point $x \in M^{n+1}$ is the orbit of x under the action of the group $\mathbb{Z}_2 = \{\pm I\}$.

Suppose that $x = (w, z)$ is an arbitrary point of $S^1 \times S^n = M^{n+1}$. Choose an orthonormal basis $\{u_1,...,u_{n+3}\}$ of $\mathbb{R}_2^{n+3}$ with u_1 and u_{n+3} timelike and the rest spacelike such that $u_1 = w$ and $u_2 = z$. Then, the tangent space

$$T_x M^{n+1} = T_w S^1 \times T_z S^n = \text{Span } \{u_3,...,u_{n+3}\}.$$

Thus, the restriction h of $<\ ,\ >$ to $T_x M^{n+1}$ has signature $(n, 1)$. The metric h is invariant under the action of $\mathbb{Z}_2$, so it induces a pseudo–Riemannian metric g of signature $(n, 1)$ on Q^{n+1}. Let π be the projection $x \to [x]$ from $\mathbb{R}_2^{n+3}$ to $\mathbb{P}^{n+2}$. Then π^*g determines a tensor field on the punctured cone $V^{n+2} - \{0\}$ which is invariant under central dilatations $x \to ax$, $a \neq 0$, and coincides with h on M^{n+1}. This metric is

(1.5) $\pi^*g\ (Y, Z) = 2 < Y, Z > /|x|^2,$

where $|x|$ is the Euclidean length of x in $\mathbb{R}^{n+3}$, and Y, Z are tangent to V^{n+2} at x. Thus, one can also consider g to be induced from the metric π^*g on the punctured cone.

The metric g can be shown to be conformally flat as follows. Let $\{u_1,...,u_{n+3}\}$ be any orthonormal basis for $\mathbb{R}_2^{n+3}$ with u_1 and u_{n+3} timelike. Let U be the open subset of points $[x]$ in Q^{n+1} whose homogeneous coordinates with respect to this basis satisfy $x_1 + x_2 \neq 0$. We will now show that U is conformally diffeomorphic to the Lorentz space $\mathbb{R}_1^{n+1}$ spanned by $\{u_3,...,u_{n+3}\}$. By taking an appropriate scalar multiple, we may assume that the homogeneous coordinates $x = (x_1,...,x_{n+3})$ of a point $[x]$ in U satisfy $x_1 + x_2 = 1$. Let $X = (x_3,...,x_{n+3})$ and let

$$(1.6) \qquad (X, X) = x_3^2 + ... + x_{n+2}^2 - x_{n+3}^2$$

be the restriction of $< , >$ to $\mathbb{R}_1^{n+1}$. Then,

$$0 = <x, x> = -x_1^2 + x_2^2 + (X, X) = -x_1 + x_2 + (X, X) ,$$

since $x_1 + x_2 = 1$. Hence, we have $x_1 - x_2 = (X, X)$, and we can solve for x_1 and x_2 as:

$$(1.7) \qquad x_1 = (1 + (X, X)) / 2 , \quad x_2 = (1 - (X, X)) / 2 .$$

Thus, we have a diffeomorphism $\beta : \mathbb{R}_1^{n+1} \to U$ defined by $\beta(X) = [\psi(X)]$, where $\psi(X)$ equals (x_1, x_2, X) for x_1 and x_2 as in (1.7).

To show that β is conformal, we consider the map $\psi : \mathbb{R}_1^{n+1} \to V^{n+2}$ and use the metric π^*g given by (1.5). Let Y be a tangent vector to $\mathbb{R}_1^{n+1}$ at the point X. From (1.7), we compute the differential $d\psi$ to be

$$d\psi(Y) = ((X, Y), - (X, Y), Y).$$

If Z is another tangent vector to $\mathbb{R}_1^{n+1}$ at X, then

$$< d\psi(Y), d\psi(Z) > = - (X, Y)(X, Z) + (X, Y)(X, Z) + (Y, Z) = (Y, Z) .$$

By (1.5) and the equation above, we have

$$\pi^*g \ (d\psi(Y), \ d\psi(Z)) = 2 \ (Y, \ Z) \ / \ |\psi(X)|^2,$$

and β is conformal.

Now we want to show that the lines in U which lie on the quadric are precisely the images under β of lightlike lines in $\mathbb{R}_1^{n+1}$. Consider two points in U with homogeneous coordinates $x = (x_1, \ x_2, \ X)$ and $y = (y_1, \ y_2, \ Y)$ satisfying the equation $x_1 + x_2 = y_1 + y_2 = 1$. The line $[x, y]$ lies on the quadric precisely when $< x, y > \ = 0$. A direct computation using (1.7) shows that

$$< x, y > \ = - (X - Y, X - Y) \ / \ 2 .$$

Hence, the line $[x, y]$ is on Q^{n+1} if and only if $X - Y$ is lightlike, i.e., the line $[X, Y]$ is a lightlike line in $\mathbb{R}_1^{n+1}$.

Since the diffeomorphism β is conformal, the paragraph above implies that lightlike vectors in the tangent space $T_q Q^{n+1}$ at any point $q \in U$ are precisely the tangent vectors to lines through q which lie on the quadric. The same can be said for all points of Q^{n+1}, since every point of the quadric lies in an open set similar to U, for an appropriate choice of homogeneous coordinate basis $\{u_1,...,u_{n+3}\}$.

We now complete the proof of Theorem 1.5. Let φ be a line–preserving diffeomorphism of Q^{n+1}. Then its differential $d\varphi$ takes lightlike vectors in the tangent space $T_q Q^{n+1}$ to lightlike vectors in the tangent space of Q^{n+1} at $\varphi(q)$. Each of these tangent spaces is isomorphic to $\mathbb{R}_1^{n+1}$. By applying Theorem 1.1 to the linear map $d\varphi$, we conclude that φ is conformal. Then, by the classification of conformal transformations of (Q^{n+1}, g) in Cahen and Kerbrat [1], φ is the restriction to Q^{n+1} of a projective transformation of $\mathbb{P}^{n+2}$ taking Q^{n+1} to itself, i.e., a Lie sphere transformation.

2.2 Generation of the Lie Sphere Group by Inversions

In this section, we will show that the group G of Lie sphere transformations and the group H of Moebius transformations are generated by inversions. This follows from the fact that the corresponding orthogonal groups are generated by reflections in hyperplanes. In fact, every orthogonal transformation on $\mathbb{R}_k^n$

is a product of at most n reflections, a result due to Cartan and Dieudonné. Our treatment of this result follows Chapter 3 of Artin's book [1]. (See also Cartan [6, pp.10–12].)

For the moment, let $< , >$ denote the scalar product of signature $(n{-}k, k)$ on $\mathbb{R}_k^n$. A hyperplane π in $\mathbb{R}_k^n$ is called *nondegenerate* if the scalar product restricts to a nondegenerate form on π. From Theorem 1.2 of Chapter 1, we know that a hyperplane π is nondegenerate if and only if its pole ξ is not lightlike. Now let ξ be a unit spacelike or timelike vector in $\mathbb{R}_k^n$. The *reflection* Ω^π of $\mathbb{R}_k^n$ in the hyperplane π with pole ξ is defined by the formula,

$$(2.1) \qquad \Omega^\pi x = x - 2\,\frac{< x,\xi >}{< \xi,\xi >}\,\xi .$$

Note that we do not define reflection in degenerate hyperplanes, i.e., those which have lightlike poles. It is clear that Ω^π fixes every point in π and that $\Omega^\pi \xi = -\,\xi$. A direct computation shows that Ω^π is in $O(n{-}k, k)$ and that $\Omega^\pi \Omega^\pi = I$.

In the proof of the theorem of Cartan and Dieudonné, we need Lemma 2.1 below concerning the special case $\mathbb{R}_k^{2k}$, where the metric has signature (k,k). In that case, let $\{e_1,...,e_{2k}\}$ be an orthonormal basis with $e_1,...,e_k$ spacelike and $e_{k+1},...,e_{2k}$ timelike. One can naturally choose a basis $\{v_1,...,v_k,w_1,...,w_k\}$ of lightlike vectors given by

$$(2.2) \qquad v_i = (e_i + e_{k+i}) / \sqrt{2} , \qquad w_i = (e_i - e_{k+i}) / \sqrt{2} .$$

Note that the scalar products of these vectors satisfy

$$(2.3) \qquad < v_i, v_j > = 0 , \qquad < w_i, w_j > = 0 , \qquad < v_i, w_j > = \delta_{ij} ,$$

for all i, j.

Let V be the lightlike subspace of dimension k spanned by $v_1,...,v_k$. Suppose that U is any other lightlike subspace of dimension k in $\mathbb{R}_k^{2k}$. Let α be any linear isomorphism of V onto U. Since both spaces are lightlike, α is trivially an isometry. By Witt's Theorem (see Artin [1, p.121]), there is an orthogonal transformation φ of $\mathbb{R}_k^{2k}$ which extends α. The vectors φv_i and φw_i satisfy the same scalar product relations (2.3) as the v_i and the w_i, and U is spanned by $\varphi v_1,...,\varphi v_k$. Using this, we can now prove the lemma.

Lemma 2.1: *Suppose that an orthogonal transformation σ fixes every vector in a lightlike subspace U of dimension k in $\mathbb{R}_k^{2k}$. Then, σ has determinant one.*

Proof: As we noted above, there exists a basis of lightlike vectors $v_1,...,v_k,w_1,...,w_k$ in $\mathbb{R}_k^{2k}$ satisfying (2.3) such that U is the subspace spanned by $v_1,...,v_k$. We now determine the matrix of σ with respect to this basis. We are given that $\sigma v_i = v_i$ for each i. Let

$$(2.4) \qquad \sigma w_j = \sum_{h=1}^{k} a_{hj} v_h + \sum_{m=1}^{k} b_{mj} w_m .$$

Since σ preserves scalar products, we have from (2.3) and (2.4),

$$b_{ij} = < v_i, \sigma w_j > = < \sigma v_i, \sigma w_j > = < v_i, w_j > = \delta_{ij} .$$

Since the matrix for σ with respect to the basis $\{v_1,...,v_k,w_1,...,w_k\}$ has zeroes below the diagonal and ones along the diagonal, σ has determinant one.   □

We now prove the theorem of Cartan and Dieudonné, following Artin [1].

Theorem 2.2: *Every orthogonal transformation of $\mathbb{R}_k^n$ is the product of at most n reflections in hyperplanes.*

Proof: The proof is by induction on the dimension n. In the case $n = 1$, the only orthogonal transformations are $\pm I$. The identity is the product of zero reflections and $-I$ is a reflection if $n = 1$. We now assume that any orthogonal transformation on a scalar product space of dimension $n - 1$ is the product of at most $n - 1$ reflections. Let σ be a given orthogonal transformation on $\mathbb{R}_k^n$. We must show that σ is the product of at most n reflections. We need to distinguish four cases.

Case 1: *There exists a non-lightlike vector v which is fixed by* σ.

In this case, let $\pi = v^{\perp}$. Since σ is orthogonal and fixes v, we have $\sigma \pi = \pi$. Let λ be the restriction of σ to π. By the induction hypothesis, λ is the product of at most $n - 1$ reflections in hyperplanes of π, say $\lambda = \Omega_1...\Omega_r$, where $r \leq n - 1$. Each Ω_i extends to a reflection in a hyperplane of $\mathbb{R}_k^n$ by setting $\Omega_i(v) = v$. Then, since σ and the product $\Omega_1...\Omega_r$ agree on π and on v, they are equal. Thus, in this case, σ is the product of at most $n - 1$ reflections.

Case 2: *There is a non-lightlike vector v such that* σ $v - v$ *is non-lightlike.*

Let π be the hyperplane $(\sigma v - v)^{\perp}$. Since σ is orthogonal, we have

$$< \sigma v + v , \sigma v - v > = < \sigma v , \sigma v > - < v , v > = 0 .$$

Thus, σ $v + v$ is in π, and we have

(2.5) $$\Omega^{\pi}(\sigma v + v) = \sigma v + v , \qquad \Omega^{\pi}(\sigma v - v) = v - \sigma v ,$$

where Ω^{π} is the reflection in the hyperplane π. Adding the two equations in (2.5) and using the linearity of Ω^{π}, we get $\Omega^{\pi} \sigma v = v$, for the non–lightlike vector v. Now by Case 1, we know that $\Omega^{\pi} \sigma$ is the product of at most $n - 1$ reflections $\Omega_1...\Omega_r$. Thus, $\sigma = \Omega^{\pi} \Omega_1...\Omega_r$ is the product of at most n reflections.

Case 3: $n = 2$.

By Cases 1 and 2, we need only handle the case where the signature of the metric is (1, 1). In that case, let $\{u, w\}$ be a basis of lightlike vectors as in (2.2) satisfying

(2.6) $$< u, u > = 0 , \qquad < w, w > = 0 , \qquad < u, w > = 1 .$$

Since σu is lightlike, it must be a multiple of u or w. We handle the two possibilities separately.

(a) Suppose that $\sigma u = a w$, for $a \neq 0$. Since σ is orthogonal, equation (2.6) implies that $\sigma w = a^{-1} u$. Then, $\sigma (u + a w) = a w + u$, and $u + a w$ is a fixed non–lightlike vector. So Case 1 applies.

(b) Suppose that $\sigma u = a u$ for $a \neq 0$. Then (2.6) implies that $\sigma w = a^{-1} w$. If the number $a = 1$, then σ is the identity, and we are done. If $a \neq 1$, we consider $v = u + w$. Then v is non–lightlike, and

$$\sigma v - v = (a - 1) u + (a^{-1} - 1) w ,$$

which is a non–lightlike vector. Hence, Case 2 applies. Thus, the only remaining case to be handled is the following.

Case 4: $n \geq 3$; *no non-lightlike vector is fixed by σ, and $\sigma v - v$ is lightlike for every non-lightlike vector v.*

Let u be any lightlike vector. We first show that $\sigma u - u$ must also be lightlike. By Theorem 1.2 of Chapter 1, we know that dim $u^\perp = n - 1$. Since $n \geq 3$, we know that $n - 1$ is greater than $n / 2$. Since the maximum possible dimension of a lightlike subspace is less than or equal to $n / 2$ by Theorem 5.1 of Chapter 1, we know that $u^\perp$ contains a non–lightlike vector v. Then, since $<v, v> \neq 0$, we have

$$< v + \varepsilon u , v + \varepsilon u > = < v , v > \neq 0 ,$$

for any real number ε. By our assumption, $\sigma v - v$ is lightlike and so also is

$$w = \sigma(v + \varepsilon u) - (v + \varepsilon u) = \sigma v - v + \varepsilon (\sigma u - u),$$

for every ε. Thus,

$$(2.7) \quad <w, w> = 2 \varepsilon <\sigma v - v , \sigma u - u> + \varepsilon^2 <\sigma u - u , \sigma u - u> = 0 ,$$

for all ε. If we take $\varepsilon = 1$ and $\varepsilon = -1$ in (2.7) and add the equations, we get

$$2 < \sigma u - u, \sigma u - u > = 0,$$

so that $\sigma u - u$ is lightlike for any lightlike vector u.

Thus, we now have that $\sigma x - x$ is lightlike for every vector x in $\mathbb{R}_k^n$. Let W be the image of the linear transformation $\sigma - I$. Then W is a lightlike subspace of $\mathbb{R}_k^n$, and so the scalar product of any two vectors in W is zero. Now let $x \in \mathbb{R}_k^n$ and $y \in W^\perp$. Then $\sigma x - x$ and $\sigma y - y$ are in W, so

$$(2.8) \quad 0 = < \sigma x - x, \sigma y - y > = < \sigma x, \sigma y > - < x, \sigma y > - < \sigma x - x, y >.$$

Since $\sigma x - x$ is in W and $y \in W^\perp$, the last term is zero. Furthermore, since $<\sigma x, \sigma y>$ equals $< x, y >$, equation (2.8) reduces to

$$< x, y - \sigma y > = 0.$$

Since this holds for all x in the scalar product space $\mathbb{R}_k^n$, we conclude that the vector $y - \sigma y = 0$, i.e., $\sigma y = y$ for all y in $W^\perp$. Since no non–lightlike vectors are fixed by σ, this implies that $W^\perp$ consists entirely of lightlike vectors. Now we have two lightlike subspaces W and $W^\perp$. We know that the sum of their dimensions is n and that each has dimension at most $n / 2$ by Theorems 1.2 and 5.1 of Chapter 1. Thus each space has dimension $n / 2$ and the signature of the metric is (k, k) where $k = n / 2$. Furthermore, since σ fixes every vector in $W^\perp$, Lemma 2.1 implies that determinant $\sigma = 1$.

Hence, the theorem holds for all orthogonal transformations of $\mathbb{R}_k^n$, with the possible exception of transformations of determinant 1 on $\mathbb{R}_k^{2k}$. In particular, any orthogonal transformation of $\mathbb{R}_k^{2k}$ with determinant -1 is the product of at most $2k$ reflections. But the product of $2k$ reflections has determinant 1, so an orthogonal transformation of $\mathbb{R}_k^{2k}$ with determinant -1 is the product of less than $2k$ reflections. Now, let Ω be any reflection in a hyperplane in $\mathbb{R}_k^{2k}$. Then $\Omega\sigma$ has determinant -1, so it is the product $\Omega_1...\Omega_r$ of less than $2k$ reflections. Therefore, $\sigma = \Omega\Omega_1...\Omega_r$ is the product of at most $2k$ reflections, as desired. □

We now return to the context of Lie sphere geometry. The Lie sphere transformation induced by a reflection Ω^π in $O(n+1,2)$ is called a *Lie inversion*. Similarly, the Moebius transformation induced by a reflection in $O(n+1,1)$ is called a *Moebius inversion*. An immediate consequence of Corollary 1.3 and Theorem 2.2 is the following.

Theorem 2.3: *The Lie sphere group G and the Moebius group H are both generated by inversions.*

In the next two sections, we will give a geometric description of these inversions and other important types of Lie sphere transformations.

2.3 Geometric Description of Inversions

In this section, we begin with a geometric description of Moebius inversions. This is followed by a more general discussion of Lie transformations, which leads naturally to the sphere geometry of Laguerre treated in the next section.

An orthogonal transformation in $O(n+1,1)$ induces a projective transformation on $\mathbb{P}^{n+1}$ which maps the Moebius sphere Σ to itself. A Moebius inversion is the projective transformation induced by a reflection Ω^π in $O(n+1,1)$. For the sake of brevity, we will also denote this projective transformation by Ω^π instead of $P(\Omega^\pi)$. Let ξ be a spacelike point in $\mathbb{P}^{n+1}$ with polar hyperplane π. The hyperplane π intersects the Moebius sphere Σ in a hypersphere S^{n-1}. The Moebius inversion Ω^π, when interpreted as a transformation on $\mathbb{R}^n$, is just ordinary inversion of $\mathbb{R}^n$ in the hypersphere S^{n-1}. We will now recall the details of this transformation.

Since the Moebius sphere is homogeneous, all inversions in planes with spacelike poles act in essentially the same way. Let us consider the special case where S^{n-1} is the sphere of radius $r > 0$ centered at the origin in $\mathbb{R}^n$. Then by formula (2.4) of Chapter 1, the spacelike point ξ in $\mathbb{P}^{n+1}$ corresponding to S^{n-1} has homogeneous coordinates

$$\xi = (\, 1 - r^2,\, 1 + r^2,\, 0 \,) / 2.$$

Let u be a point in $\mathbb{R}^n$ other than the origin. By equation (2.1) of Chapter 1, the point u corresponds to the point in $\mathbb{P}^{n+1}$ with homogeneous coordinates

$$x = (1 + u \cdot u , 1 - u \cdot u , 2u) / 2 .$$

The formula for $\Omega^\pi x$ in homogeneous coordinates is

(3.1) $$\Omega^\pi x = x - \frac{2(x,\xi)}{(\xi,\xi)} \xi ,$$

where $(\ ,\ )$ is the Lorentz metric. A straightforward calculation shows that $\Omega^\pi x$ is the point in $\mathbb{P}^{n+1}$ with homogeneous coordinates

$$(1 + v \cdot v , 1 - v \cdot v , 2v) / 2 ,$$

where $v = (r^2/|u|^2) u$. Thus, the Euclidean transformation induced by Ω^π maps u to the point v on the ray through u from the origin satisfying the equation $|u|\,|v| = r^2$. From this, it is clear that the fixed points of Ω^π are precisely the points of the sphere S^{n-1}. Viewed in the projective context, this is immediately clear from (3.1), since $\Omega^\pi x = x$ if and only if $(x, \xi) = 0$. In general, inversion of $\mathbb{R}^n$ in the hypersphere of radius r centered at a point p maps a point $u \neq p$ to the point v on the ray through u from p satisfying

$$| u - p | \, | v - p | = r^2 .$$

Another special case is when the unit spacelike vector ξ lies in the Euclidean space $\mathbb{R}^n$ spanned by $e_3,...,e_{n+2}$. Then the "sphere" corresponding to $[\xi]$ by formula (2.6) of Chapter 1 is the hyperplane V through the origin in $\mathbb{R}^n$ perpendicular to ξ. In this case, the Moebius inversion Ω^π is just ordinary Euclidean reflection of $\mathbb{R}^n$ in the hyperplane V.

It is also instructive to study inversion as a map from the Moebius sphere Σ to itself. Suppose that ξ is a spacelike point in $\mathbb{P}^{n+1}$ and x is a point on Σ which is not on the polar hyperplane π of ξ. The line $[x, \xi]$ in $\mathbb{P}^{n+1}$ intersects Σ in precisely two points, x and $\Omega^\pi x$ (see Figure 3.1), and Ω^π simply exchanges these two points.

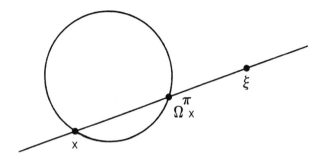

Figure 3.1 – Inversion Ω^{π} with spacelike pole ξ

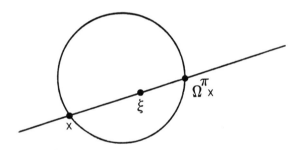

Figure 3.2 – Inversion Ω^{π} with timelike pole ξ

Given a spacelike point η , the sphere polar to η is taken by Ω^{π} to the sphere polar to $\Omega^{\pi}\eta$, since Ω^{π} is an orthogonal transformation. Thus, the sphere polar to η is taken to itself by Ω^{π} if and only if η is fixed by Ω^{π}, i.e., $(\xi, \eta) = 0$. Geometrically, this means that the sphere polar to η is orthogonal to the sphere of inversion.

If ξ is a timelike point in $\mathbb{P}^{n+1}$, then formula (3.1) still makes sense, although the polar hyperplane π to ξ does not intersect Σ, and thus $(x, \xi) \neq 0$ for each x in Σ. Given a point x in Σ, the line $[x, \xi]$ intersects Σ in precisely two points (see Figure 3.2), and Ω^{π} interchanges these two points. Suppose, for example, we take $\xi = e_1$ and represent a point on Σ by homogeneous coordinates $(1, y)$, where y is a vector in the span of $e_2,...,e_{n+2}$ satisfying the equation $y \cdot y = 1$, i.e., $y \in S^n$. Then from (3.1), we see that

$$\Omega^{\pi} (1, y) = (-1, y) \cong (1, -y) ,$$

and Ω^{π} acts as the antipodal map on the sphere S^n. Note that the projective transformation Ω^{π} equals the projective transformation induced by the product $\Omega_2...\Omega_{n+2}$, where Ω_j is inversion in the polar hyperplane $e_j^{\perp}$, because the corresponding orthogonal transformations differ by a minus sign. This is true independent of the choice of hyperplane π with timelike pole. Therefore, we have established the following refinement of Theorem 2.3 for the Moebius group.

Theorem 3.1: *The Moebius group H is generated by inversions in polar hyperplanes to spacelike points of $\mathbb{P}^{n+1}$, i.e., by inversions in spheres in Σ.*

We now return to the setting of Lie geometry of oriented spheres. In general, it is hard to give a geometric description of a Lie inversion which is not induced by a Moebius inversion. One noteworthy special case, however, is the change of orientation transformation Γ (see Remark 1.4) which is the Lie inversion Ω^{π} determined by the hyperplane π orthogonal to e_{n+3} .

We next present an alternative way to view the Lie sphere group G by decomposing it into certain natural subgroups. To do this, we need the concept of a linear complex of spheres. The *linear complex of spheres* determined by a point ξ in $\mathbb{P}^{n+2}$ is the set of all spheres represented by points x in the Lie quadric Q^{n+1} satisfying the equation $< x, \xi > = 0$. The complex is called:

> *elliptic* if ξ is spacelike,
>
> *hyperbolic* if ξ is timelike,
>
> *parabolic* if ξ is lightlike.

Since the Lie sphere group G acts transitively on each of the three types of points, each linear complex of a given type looks like every other complex of the same type. A typical example of an elliptic complex is obtained by taking $\xi = e_{n+2}$. A sphere S in $\mathbb{R}^n$ represented by a point x on Q^{n+1} satisfies the equation $< x, \xi > = 0$ if and only if its coordinate $x_{n+2} = 0$, i.e., the center of S lies in the hyperplane $\mathbb{R}^{n-1}$ with equation $x_{n+2} = 0$ in $\mathbb{R}^n$. The linear complex consists of all spheres and planes orthogonal to this plane, including the points of the plane as a special case. A Lie sphere transformation T maps each sphere in the complex to another sphere in the complex if and only if $e_{n+2}^{\perp}$ is an invariant subspace of T. Since T can be represented by an orthogonal transformation, this is equivalent to $T[e_{n+2}] = [e_{n+2}]$. Thus, T is determined by its action on $e_{n+2}^{\perp}$. Let $\mathbb{R}_2^{n+2}$ denote the vector subspace $e_{n+2}^{\perp}$ in $\mathbb{R}_2^{n+3}$ endowed with the metric $< , >$ inherited from $\mathbb{R}_2^{n+3}$, and let $O(n,2)$ denote the group of orthogonal transformations of $\mathbb{R}_2^{n+2}$. A transformation A in $O(n,2)$ can be extended to $\mathbb{R}_2^{n+3}$ by setting $Ae_{n+2} = e_{n+2}$. This gives an isomorphism between $O(n,2)$ and the group of Lie transformations which fix the elliptic complex. This group is a double covering of the group of Lie transformations of the Euclidean space $\mathbb{R}^{n-1}$ orthogonal to e_{n+2} in $\mathbb{R}^n$.

A typical example of a hyperbolic complex is the case $\xi = e_{n+3}$. This complex consists of all point spheres. A second example is the complex corresponding to $\xi = (-r, r, 0, \ldots, 0, 1)$. This complex consists of all oriented spheres with signed radius r. The group of Lie sphere transformations which map this hyperbolic complex to itself consists of all transformations which map the projective point ξ to itself. This group is isomorphic to the Moebius subgroup of G, as discussed in Remark 1.4.

The parabolic complex determined by a point ξ on Q^{n+1} consists of all spheres in oriented contact with the sphere corresponding to ξ. A noteworthy example is the case $\xi = (1,-1,0,...,0)$, the improper point. This system consists of all oriented hyperplanes in $\mathbb{R}^n$. A Lie sphere transformation which fixes this complex is called a Laguerre transformation, and the group of Laguerre transformations is called the Laguerre group. We will study this group in detail in the next section.

2.4 Laguerre Geometry

Each point in the intersection of the Lie quadric with the plane $x_1 + x_2 = 0$ represents either a plane in $\mathbb{R}^n$ or the improper point. The other points in the quadric represent actual spheres in $\mathbb{R}^n$, including point spheres. The homogeneous coordinates of points in this complementary set satisfy the condition $x_1 + x_2 \neq 0$. The following elementary lemma shows that a Lie transformation is determined by its action on such points.

Lemma 4.1: *A Lie sphere transformation is determined by its restriction to the set of points $[x]$ in Q^{n+1} with $x_1 + x_2 \neq 0$.*

Proof: To prove this, it is sufficient to exhibit a basis of lightlike vectors in $\mathbb{R}_2^{n+3}$ satisfying $x_1 + x_2 \neq 0$. One can check that $v_1,...,v_{n+3}$ given below is such a basis:

$$v_1 = e_2 + e_{n+3}, \qquad v_i = e_1 + e_i, \quad 2 \leq i \leq n+2, \qquad v_{n+3} = e_3 - e_{n+3}. \quad \square$$

We now show that the set of points in Q^{n+1} with $x_1 + x_2 \neq 0$ is naturally diffeomorphic to the Lorentz space $\mathbb{R}_1^{n+1}$ spanned by $e_3,...,e_{n+3}$. By taking the appropriate scalar multiple, we may assume that the homogeneous coordinates of the point $[x]$ satisfy $x_1 + x_2 = 1$. Let $X = (x_3,..., x_{n+3})$, and let

(4.1) $$(X, X) = x_3^2 + ... + x_{n+2}^2 - x_{n+3}^2,$$

denote the restriction of $< , >$ to $\mathbb{R}_1^{n+1}$. Then,

$$0 = \,<x, x> \,= -x_1^2 + x_2^2 + (X, X) = -x_1 + x_2 + (X, X),$$

since $x_1 + x_2 = 1$. Hence, we have $x_1 - x_2 = (X, X)$, and we can solve for x_1 and x_2 as follows:

$$(4.2) \qquad x_1 = (1 + (X, X)) / 2, \quad x_2 = (1 - (X, X)) / 2.$$

Thus, we have a diffeomorphism $[x] \to X$ between the open set U of points in Q^{n+1} with $x_1 + x_2 \neq 0$ and points X in $\mathbb{R}_1^{n+1}$. In the proof of Theorem 1.5, it was shown that this diffeomorphism is conformal if Q^{n+1} is endowed with the standard pseudo–Riemannian metric (see Cahen and Kerbrat [1, p.327]). From formula (3.4) of Chapter 1, we see that the center p and signed radius r of the sphere determined by X are given by

$$(4.3) \qquad p = (x_3,...,x_{n+2}), \quad r = x_{n+3}.$$

Geometrically, one obtains the sphere S determined by X as the intersection of the plane $x_{n+3} = 0$ with the light (isotropy) cone with vertex X. The orientation of the sphere is determined by the sign of x_{n+3}. The mapping which takes X to the oriented sphere obtained this way was classically called *isotropy projection* (see Figure 4.1 and Blaschke [1, p.136]). Note that the spheres corresponding to two points X and Y in $\mathbb{R}_1^{n+1}$ are in oriented contact if and only if the line determined by X and Y is lightlike (see Figure 4.2). To see this analytically, suppose that $x = (x_1, x_2, X)$ and $y = (y_1, y_2, Y)$ are the homogeneous coordinates of two points on the quadric satisfying

$$<x, x> \,= \,<y, y> \,= 0, \quad x_1 + x_2 = y_1 + y_2 = 1.$$

The spheres corresponding to $[x]$ and $[y]$ are in oriented contact if and only if $<x, y> \,= 0$. Using (4.2) for x_1, x_2, y_1 and y_2, a direct computation shows that

$$(4.4) \qquad <x, y> \,= -x_1 y_1 + x_2 y_2 + (X, Y) = -(X - Y, X - Y) / 2.$$

Thus, the spheres are in oriented contact if and only if the vector $X - Y$ is

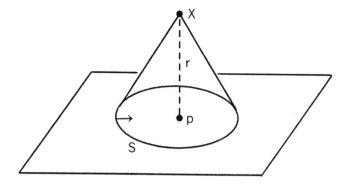

Figure 4.1 – Isotropy projection

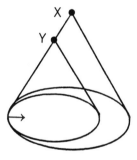

Figure 4.2 – Oriented contact and isotropy projection

lightlike. From this, it is obvious that a parabolic pencil of spheres in $\mathbb{R}^n$ corresponds to a lightlike line in $\mathbb{R}^{n+1}_1$.

It is also possible to represent oriented hyperplanes of $\mathbb{R}^n$ in Laguerre geometry. The idea is to identify an oriented hyperplane π having unit normal N with the set of all contact elements (p, N), where p is a point of π. By isotropy projection, the parabolic pencil of spheres in oriented contact at (p, N) corresponds to a lightlike line in $\mathbb{R}^{n+1}_1$. The union of all these lightlike lines is an isotropy plane, i.e., an affine hyperplane V in $\mathbb{R}^{n+1}_1$ whose pole with respect to the Lorentz metric is lightlike. Such a plane meets $\mathbb{R}^n$ at an angle of $\pi/4$ (see Figure 4.3). Thus, we have a bijective correspondence between oriented planes in $\mathbb{R}^n$ and isotropy planes in $\mathbb{R}^{n+1}_1$.

A fundamental geometric quantity in Laguerre geometry is the tangential distance between two spheres. To study this, we first need to resolve the question of when two oriented spheres in $\mathbb{R}^n$ have a common tangent (oriented) plane. While it is obvious that some pairs of spheres have a common tangent plane (see Figure 4.4), it is just as obvious that some pairs, such as two concentric spheres, do not. The following lemma answers this question.

Lemma 4.2: *The two oriented spheres in $\mathbb{R}^n$ corresponding to the points X and Y in $\mathbb{R}^{n+1}_1$ have a common tangent plane if and only if $X - Y$ is lightlike or spacelike.*

Proof: From the discussion above, we know that the two spheres corresponding to X and Y have a common tangent plane if and only if there is an isotropy plane π in $\mathbb{R}^{n+1}_1$ which contains X and Y, i.e., there is a lightlike vector v such that $(X - Y, v) = 0$. First suppose that $X - Y$ is timelike. Then the Lorentz metric is positive definite on the orthogonal complement of $X - Y$, and there is no isotropy plane which contains X and Y. Next, if $X - Y$ is lightlike, then the corresponding spheres are in oriented contact at a certain contact element (p, N), and they have exactly one common tangent plane determined by (p, N). Finally, suppose that $X - Y$ is spacelike. Then the Lorentz metric has signature $(n-1, 1)$ on $(X - Y)^\perp$. Let v be any lightlike vector orthogonal to $X - Y$. Then the line $[X, Y]$ lies in the isotropy plane π through X with pole v, and so the spheres determined by X and Y have a

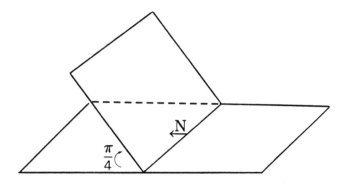

Figure 4.3 – Isotropy plane

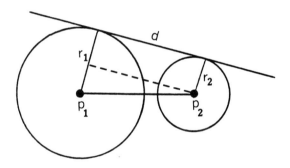

Figure 4.4 – Tangential distance between spheres

common tangent plane corresponding to the intersection of π with $\mathbb{R}^n$. This shows that the set of common tangent planes to the two spheres is in bijective correspondence with the set of projective classes of lightlike vectors in the projective space $\mathbb{P}^{n-1}$ determined by the n–plane $(X - Y)^\perp$. As we saw in Section 1.1, this set is naturally diffeomorphic to an $(n-2)$–sphere.   □

Suppose that $X = (p_1, r_1)$ and $Y = (p_2, r_2)$ are two points representing spheres S_1 and S_2 which have a common tangent plane (see Figure 4.4). By symmetry, it is clear that all common tangent segments to the two spheres have the same Euclidean length. This length d is called the *tangential distance* between S_1 and S_2. By constructing a right triangle as in Figure 4.4, we see that

$$d^2 + |r_1 - r_2|^2 = |p_1 - p_2|^2 \,,$$

and thus,

(4.5) $$d^2 = |p_1 - p_2|^2 - |r_1 - r_2|^2 = (X - Y, X - Y)\,.$$

Hence, the tangential distance is just the square root of the non–negative quantity $(X-Y, X-Y)$. Of course, the tangential distance is zero precisely when the two spheres are in oriented contact.

A Lie sphere transformation which maps the improper point to itself is called a *Laguerre transformation*. Since oriented contact must be preserved, Laguerre transformations can also be characterized as those Lie transformations which take planes to planes. Consequently, a Laguerre transformation maps the open set U determined by $x_1 + x_2 \neq 0$ onto itself. Through the correspondence between U and $\mathbb{R}_1^{n+1}$ given via equation (4.2), a Laguerre transformation induces a transformation of $\mathbb{R}_1^{n+1}$ onto itself. We will now show that this must be an affine transformation.

Suppose that $\alpha = P(\sigma)$ is the Laguerre transformation determined by a transformation $\sigma \in O(n+1,2)$. As a transformation of $\mathbb{R}_2^{n+3}$, σ takes the affine plane π given by $x_1 + x_2 = 1$ to another affine plane in $\mathbb{R}_2^{n+3}$. If $\sigma\pi$ were not parallel to π, then $\sigma\pi$ would intersect the plane $x_1 + x_2 = 0$, contradicting the assumption that α maps U to U. Thus, $\sigma\pi$ is given by the equation $x_1+x_2 = c$, for some $c \neq 0$. If $A = \sigma / c$, then A induces the same Lie transformation α as σ, but A takes the plane π to itself. Thus, we now represent α by the transformation A, which is not in $O(n+1,2)$ unless $c = \pm 1$. Suppose that the transformation A is represented by the matrix

$$A = [\, a_{ij} \,] , \quad 1 \leq i, j \leq n+3,$$

with respect to the standard basis of $\mathbb{R}_2^{n+3}$. Since α takes the improper point to itself, we have

$$A(1,-1,0,...,0) = \lambda \, (1,-1,0,...,0) ,$$

for some $\lambda \neq 0$. From matrix multiplication and the equation above, we get

(4.6) $$a_{i1} = a_{i2}, \quad 3 \leq i \leq n+3 .$$

Suppose now that $x = (x_1,...,x_{n+3})$ with $x_1 + x_2 = 1$, and let $Ax = y = (y_1,...,y_{n+3})$. By our choice of A, we know that $y_1 + y_2 = 1$ also. Thus, x and y are determined by $X = (x_3,...,x_{n+3})$ and $Y = (y_3,...,y_{n+3})$ using formula (4.2). Since $a_{i1} = a_{i2}$ and $x_1 + x_2 = 1$, we have

(4.7) $$y_i = a_{i1}x_1 + a_{i2}x_2 + \sum_{j=3}^{n+3} a_{ij}x_j = a_{i1} + \sum_{j=3}^{n+3} a_{ij}x_j .$$

Thus, Y is obtained from X by the affine transformation T given by

(4.8) $$Y = TX = BX + C ,$$

where B is the invertible linear transformation of $\mathbb{R}_1^{n+1}$ represented by the matrix $[\, a_{ij} \,]$, $3 \leq i, j \leq n+3$, and C is the vector $(a_{31},...,a_{n+3 \, 1})$. The fact that this transformation must preserve oriented contact of spheres implies further that T must preserve the relationship

(4.9) $(X - Z, X - Z) = 0$.

The discussion above suggests the possibility of simply beginning in the Laguerre space $\mathbb{R}_1^{n+1}$, as in Blaschke [1, p.136]. From that point of view, an affine transformation $TX = BX + C$ of $\mathbb{R}_1^{n+1}$ which preserves the relationship (4.9) is called an *affine Laguerre transformation*. If the vector C is zero, then T is called a *linear Laguerre transformation*. Of course, the linear part B must be invertible by definition of an affine transformation.

We have seen that a Lie transformation which takes the improper point to itself induces an affine Laguerre transformation. Conversely, we will show that an affine Laguerre transformation extends in a unique way to a Lie transformation. Before that, however, we want to establish some important properties of the group of affine Laguerre transformations. Theorem 1.1 and the assumption that (4.9) is preserved yield the following result concerning the linear part of an affine Laguerre transformation.

Theorem 4.3: *Suppose that $TX = BX + C$ is an affine Laguerre transformation. Then $B = \mu\, D$, where $D \in O(n,1)$ and $\mu > 0$.*

Proof: Suppose that X and Z are any two points in $\mathbb{R}_1^{n+1}$ such that $X - Z$ is lightlike. Then, since T preserves (4.9), we have

$$0 = ((BX{+}C) - (BZ{+}C), (BX{+}C) - (BZ{+}C)) = (B(X{-}Z), B(X{-}Z)) .$$

Thus, the linear transformation B takes lightlike vectors to lightlike vectors. By Theorem 1.1 and our standing assumption that $n \geq 2$, we know that there exists a positive number λ such that $(Bv, Bw) = \lambda\, (v, w)$, for all v, w in $\mathbb{R}_1^{n+1}$. Then $D = \lambda^{-1/2} B$ is orthogonal, and we have $B = \mu D$ for $\mu = \lambda^{1/2}$.   □

We now want to give a geometric interpretation of this decomposition of B. First, it is immediate from (4.5) and Theorem 4.3 that the orthogonal transformations are precisely those linear Laguerre transformations which preserve tangential distances between spheres. Secondly, consider the linear transformation $S^\mu = \mu\, I$, where $\mu > 0$ and I is the identity transformation on

$\mathbb{R}_1^{n+1}$. This transformation takes the point (p, r) to $(\mu p, \mu r)$. When interpreted as a map on the space of spheres, it takes the sphere with center p and signed radius r to the sphere with center μp and signed radius μr. Thus, S^μ is one of the two affine Laguerre transformations induced from the Euclidean central dilatation $p \rightarrow \mu p$, for $p \in \mathbb{R}^n$. The transformation S^μ preserves the sign of the radius and hence the orientation of each sphere in $\mathbb{R}^n$. The other affine Laguerre transformation induced from the same central dilatation is ΓS^μ, where Γ is the change of orientation transformation. Thus, we have the following geometric version of Theorem 4.3.

Corollary 4.4: *Every linear Laguerre transformation B can be written in the form $B = S^\mu D$, where S^μ is the orientation preserving Laguerre transformation induced by a central dilatation of $\mathbb{R}^n$, and D preserves tangential distances between spheres.*

Next, we consider the effect of a *Laguerre translation* $TX = X + C$ on the space of spheres. Suppose first that $C = (v, 0)$ for $v \in \mathbb{R}^n$. Then we have $T(p, r) = (p + v, r)$, so T translates the center of every sphere by the vector v while preserving the signed radius. We see that T is just the orientation preserving affine Laguerre transformation induced from the Euclidean translation $p \rightarrow p+v$. We will denote T by τ_v.

Now consider the case where $C = (0, t)$, $t \in \mathbb{R}$. Then $T(p, r) = (p, r + t)$, and T adds t to the signed radius of every sphere while keeping the center fixed. T is called *parallel transformation* by t and will be denoted P_t. Note that P_t is a Laguerre transformation which is not a Moebius transformation. It takes point spheres to spheres with radius t and takes spheres of radius $-t$ to point spheres. Thus, the group of Laguerre translations is a commutative subgroup of the group of affine Laguerre transformations, and we have shown that it decomposes as follows.

Theorem 4.5: *Any Laguerre translation T can be written in the form $T = P_t \tau_v$, where P_t is a parallel transformation and τ_v is the orientation preserving Laguerre translation induced by Euclidean translation by the vector v.*

In studying the role of inversions in the group of affine Laguerre transformations, we must consider inversions in planes in $\mathbb{R}^{n+1}_1$ which do not contain the origin. Let π be an affine plane in $\mathbb{R}^{n+1}_1$ whose pole ξ is not lightlike. Then the inversion Ω^π of $\mathbb{R}^{n+1}_1$ in π is defined by the formula

(4.10) $\Omega^\pi X = X - (\, 2(X - P, \xi) / (\xi, \xi) \,)\, \xi\,,$

where P is any point on π. We will call Ω^π a *Laguerre inversion*. It is easy to see that the group of affine Laguerre transformations is generated by Laguerre inversions along with the transformations S^μ induced by central dilatations. First, we know from Theorem 4.3 that any linear Laguerre transformation is the product of some S^μ with an orthogonal transformation D. An orthogonal transformation D is the product of Laguerre inversions by Theorem 2.2. As for translations, it is well known that Euclidean translation by a vector v is the product of two Euclidean reflections $\Omega_2\Omega_1$ in parallel planes π_1 and π_2 orthogonal to v such that $v / 2$ is the vector from any given point P on π_1 to its closest point Q on π_2 . (See Ryan [2, p.23] for more detail.) The Laguerre translation τ_v is the product of the Laguerre inversions induced by these two Euclidean reflections.

To express a parallel transformation P_t as the product of two inversions, we first note that the change of orientation transformation Γ is a Laguerre inversion in the plane π_1 through the origin with pole e_{n+3} . If Ω_2 is the Laguerre inversion in the plane π_2 through the point $(0, t/2)$ with pole e_{n+3} , then P_t is the product $\Omega_2\Gamma$.

Finally, we want to show that every affine Laguerre transformation extends to a Lie transformation. Such an extension is necessarily unique by Lemma 4.1. First, recall that if α is the Lie transformation extending an affine Laguerre transformation $TX = BX + C$, then α has a unique representative A in $GL(n+3)$ which takes the plane $x_1 + x_2 = 1$ to itself. The matrix $[\, a_{ij}\,]$ for A with respect to the standard basis is largely determined by B and C through (4.8), i.e.,

(4.11) $[B] = [\, a_{ij}\,]\,,\quad 3 \le i, j \le n+3\,,\quad C = (a_{31}, \ldots, a_{n+3\ 1})\,.$

We now show how to determine the rest of A.

First, consider the case where B is a linear Laguerre transformation. The Lie extension of B must take the improper point $[e_1 - e_2]$ to itself and the point $[e_1 + e_2]$ corresponding to the origin in R_1^{n+1} to itself. This means that the transformation A satisfies

(4.12) $A(e_1 - e_2) = a \, (e_1 - e_2), \quad A(e_1 + e_2) = b \, (e_1 + e_2) \, ,$

for some non–zero scalars a and b. From (4.12) and the linearity of A, we obtain

(4.13) $Ae_1 = c \, e_1 + d \, e_2 , \quad Ae_2 = d \, e_1 + c \, e_2 ,$

where $c = (a+b) \, / \, 2$ and $d = (b-a) \, / \, 2$. Thus, the span R_1^2 of e_1 and e_2 is an invariant subspace of A. Since A is a scalar multiple of an orthogonal transformation, the orthogonal complement R_1^{n+1} of R_1^2 is also invariant under A. Therefore, the matrix for A has the form

(4.14) $A = \begin{bmatrix} J & 0 \\ 0 & B \end{bmatrix} \quad \text{where} \quad J = \begin{bmatrix} c & d \\ d & c \end{bmatrix} .$

By Corollary 4.4, every linear Laguerre transformation is of the form $S^\mu D$, where S^μ is induced from central dilatation of R^n and $D \in O(n,1)$. We now show how to extend each of these two types of transformations. First, consider the case of a linear Laguerre transformation determined by D in $O(n,1)$. The Lie extension of D is obtained by taking $c = 1$, $d = 0$ and $B = D$ in (4.14). To check this, let X be an arbitrary point in R_1^{n+1}. By (4.2), X corresponds to the point in Q^{n+1} with homogeneous coordinates

(4.15) $x = (\, 1 + (X, X) \, , \, 1 - (X, X) \, , \, 2X \,) \, / \, 2 \, .$

The point DX in R_1^{n+1} corresponds to the point in Q^{n+1} with coordinates

$y = (\, 1 + (DX, DX) \, , \, 1 - (DX, DX), \, 2 \, DX \,) \, / \, 2 \, .$

However, since $(DX, DX) = (X, X)$, we have

$$y = (1 + (X, X) , 1 - (X, X) , 2\, DX) / 2 .$$

It is now clear that $y = Ax$, where A is in the form (4.14) with $c = 1$, $d = 0$ and $B = D$. Note that this matrix is in $O(n+1,2)$.

Next, consider the linear Laguerre transformation $S^{\mu} = \mu\, I$ on $\mathbb{R}_1^{n+1}$. Let A have the form (4.14) with $B = \mu\, I$. Let $X \in \mathbb{R}_1^{n+1}$ and x be as in (4.15). By matrix multiplication, the first two coordinates of $y = Ax$ are given by

(4.16) $y_1 = ((c+d) + (c-d)\,(X, X)) / 2 , \quad y_2 = ((c+d) + (d-c)\,(X, X)) / 2 .$

On the other hand, the point μX corresponds to the point in Q^{n+1} with coordinates

(4.17) $(1 + \mu^2\,(X, X) , 1 - \mu^2\,(X, X) , 2\,\mu X) / 2 .$

Equating y_1 and y_2 with the first two coordinates of (4.17) yields

(4.18) $c = (1 + \mu^2) / 2 , \quad d = (1 - \mu^2) / 2 .$

The matrix A in (4.14) with these values of c and d and $B = \mu\, I$ is the extension of S^{μ}. Note that the numbers c and d in (4.18) satisfy $c^2 - d^2 = \mu^2$. Since $\mu > 0$, the number c / μ is positive, and so there exists a real number t such that

(4.19) $c / \mu = \cosh t, \quad d / \mu = \sinh t .$

From (4.18) and (4.19) one can determine that $t = -\ln \mu$. If we divide the matrix A by μ, we obtain the following orthogonal matrix which also represents the Lie transformation extending S^{μ},

(4.20) $\begin{bmatrix} K & 0 \\ 0 & I \end{bmatrix}$ where $K = \begin{bmatrix} \cosh t & \sinh t \\ \sinh t & \cosh t \end{bmatrix} .$

Finally, we turn to the problem of extending the parallel transformation P_t. The linear part B of P_t is the identity I, while the translation part is the vector $C = (0, t)$. By (4.11), this determines much of the matrix A of the extension of P_t. Further, since the improper point is mapped to itself,

$$(4.21) \qquad A(1,-1,0,...,0) = a\,(1,-1,0,...,0)\,,$$

for some $a \neq 0$. Since P_t takes the point $(0, 0)$ in $\mathbb{R}_1^{n+1}$ to $(0, t)$, the transformation A must satisfy

$$(4.22) \qquad A(1,1,0,...,0) = b\,(\,1 - t^2,\,1 + t^2,\,0\,,\,2t\,)\,/\,2\,,$$

for some $b \neq 0$. By adding (4.21) and (4.22), we can determine $A(e_1)$. Since $a_{n+3\,1} = t$ by (4.11), b must equal 2. From the results so far, we know that the subspace $V = \text{Span}\,\{e_1, e_2, e_{n+3}\}$ is invariant under A. Therefore, $V^{\perp}$ is also invariant. This and the fact that $B = I$ imply that $Ae_i = e_i$ for $3 \leq i \leq n+2$. Since A magnifies all vector lengths by the same amount, this implies that A is orthogonal. From these orthogonal relations, one can determine that $a = 1$ and ultimately that A has the form below. For future reference, we will now denote this transformation by P_t instead of A,

$$(4.23) \qquad P_t = \begin{bmatrix} 1-(t^2/2) & -t^2/2 & 0 \dots 0 & -t \\ t^2/2 & 1+(t^2/2) & 0 \dots 0 & t \\ 0 & 0 & & \\ & & & I \\ t & t & & \end{bmatrix}.$$

2.5 Subgeometries of Lie Sphere Geometry

We close this chapter by examining some important subgeometries of Lie sphere geometry from the point of view of Klein's Erlangen Program. These are the geometries of Moebius and Laguerre, and the Euclidean, spherical and hyperbolic metric geometries. By making use of the concept of a Legendre submanifold, to be introduced in the next chapter, one can study submanifold theory in any of these subgeometries within the context of Lie sphere geometry.

The subgroup of Moebius transformations consists of those Lie transformations which take point spheres to point spheres. These are precisely the Lie transformations which take the point $[e_{n+3}]$ to itself. As we saw in Remark 1.4, this Moebius group is isomorphic to $O(n+1,1)$.

The subgroup of Laguerre transformations consists of those Lie transformations which take planes to planes in $\mathbb{R}^n$. These are the Lie transformations which take the improper point $[e_1 - e_2]$ to itself. In the preceding section, we saw that each Laguerre transformation corresponds to an affine Laguerre transformation of the space $\mathbb{R}^{n+1}_1$ spanned by $e_3,...,e_{n+3}$.

As before, let $\mathbb{R}^n$ denote the Euclidean space spanned by $e_3,...,e_{n+2}$. Recall that a *similarity transformation* of $\mathbb{R}^n$ is a mapping φ from $\mathbb{R}^n$ to itself, such that for all p and q in $\mathbb{R}^n$, the Euclidean distance $d(p, q)$ is transformed as follows,

$$d(\varphi p, \varphi q) = \kappa \, d(p, q) ,$$

for some constant $\kappa > 0$. Every similarity transformation can be written as a central dilatation followed by an isometry of $\mathbb{R}^n$. The group of Lie transformations induced by similarity transformations is clearly a subgroup of both the Laguerre group and the Moebius group. The next theorem shows that it is precisely the intersection of these two subgroups.

Theorem 5.1: (a) *The intersection of the Laguerre group and the Moebius group is the group of Lie transformations induced by similarity transformations of* $\mathbb{R}^n$.
(b) *The group G of Lie transformations is generated by the union of the groups of Laguerre and Moebius.*

Proof: (a) By the results of the last section, the Laguerre group is isomorphic to the group of affine Laguerre transformations on $\mathbb{R}^{n+1}_1$. An affine Laguerre transformation $TX = BX + C$ is also a Moebius transformation if and only if $Te_{n+3} = \pm e_{n+3}$. Since $T(0) = C$, this immediately implies that $C = (v, 0)$, for some v in $\mathbb{R}^n$. Next, by Corollary 4.4, the linear part B of T is of the form

$S^\mu D$, where S^μ is induced from a central dilatation of $\mathbb{R}^n$ and D is in $O(n,1)$. Since S^μ is a Moebius transformation, T is a Moebius transformation precisely when D is Moebius, i.e., $De_{n+3} = \pm e_{n+3}$. This means that the matrix for D with respect to the standard basis of $\mathbb{R}_1^{n+1}$ has the form

(5.1)
$$D = \begin{bmatrix} A & 0 \\ 0 & \pm 1 \end{bmatrix} , \quad A \in O(n).$$

Thus, D is one of the two Laguerre transformations induced by the linear isometry A of $\mathbb{R}^n$, and B is a similarity transformation. Then T is also a similarity transformation, and (a) is proven.

(b) Let α be a Lie transformation. If $\alpha [e_1 - e_2] = [e_1 - e_2]$, then α is a Laguerre transformation. Next, suppose that $\alpha [e_1 - e_2]$ is a point $[x]$ in Q^{n+1} with $x_1 + x_2 = 0$. Then $\alpha [e_1 - e_2]$ corresponds to a plane π in $\mathbb{R}^n$. Let I_1 be an inversion in a sphere whose center is not on the plane π. Then $[y] = I_1 [x]$ is a point in Q^{n+1} with $y_1 + y_2 \neq 0$, i.e., $[y]$ corresponds to a sphere in $\mathbb{R}^n$. Let p and r denote the center and signed radius of this sphere. If $\alpha [e_1 - e_2]$ does not correspond to a plane, then the step above is not needed. In that case, we let I_1 be the identity transformation.

Next, the parallel transformation P_{-r} takes the point $[y]$ to the point $[z]$ with $z_{n+3} = 0$ corresponding to the point sphere $p \in \mathbb{R}^n$. Finally, an inversion I_2 in a sphere centered at p takes $[z]$ to the improper point $[e_1 - e_2]$. Since the transformation $I_2 P_{-r} I_1 \alpha$ takes $[e_1 - e_2]$ to itself, it is a Laguerre transformation Ψ. Since each inversion is its own inverse and the inverse of P_{-r} is P_r , we have $\alpha = I_1 P_r I_2 \Psi$, a product of Laguerre and Moebius transformations.   □

From the proof of Theorem 5.1(a), we have the following immediate corollary.

Corollary 5.2: *The group of Lie transformations induced from isometries of $\mathbb{R}^n$ is isomorphic to the set of affine Laguerre transformations $TX = DX + C$, where D has the form (5.1) and $C = (v, 0)$, for $v \in \mathbb{R}^n$.*

Two other important subgeometries of Moebius geometry are the metric geometries of the sphere S^n and hyperbolic space H^n (see Section 1.4). Let S^n

be the unit sphere in the Euclidean space $\mathbb{R}^{n+1}$ spanned by $e_2,...,e_{n+2}$. The group of isometries of S^n is the orthogonal group $O(n+1)$ of linear transformations which preserve the Euclidean metric on $\mathbb{R}^{n+1}$. An isometry A of S^n induces a Moebius transformation whose matrix with respect to the standard basis $e_1,...,e_{n+2}$ of $\mathbb{R}_1^{n+2}$ is

(5.2) $$\begin{bmatrix} 1 & 0 \\ 0 & A \end{bmatrix} .$$

The group of isometries of S^n is clearly isomorphic to the subgroup of Moebius transformations of this form. Of course, each such Moebius transformation induces two Lie transformations differing by Γ.

As in Section 1.4, we represent hyperbolic space H^n by the set of points y in the Lorentz space $\mathbb{R}_1^{n+1}$ spanned by $e_1,e_3,...,e_{n+2}$ satisfying $(y, y) = -1$, where $y_1 \geq 1$. (Note that this is a different Lorentz space than the one used in Laguerre geometry.) H^n is one of the two components of the subset of $\mathbb{R}_1^{n+1}$ on which $(y, y) = -1$. The group of isometries of H^n is a subgroup of index 2 in the orthogonal group $O(n,1)$ of linear isometries of $\mathbb{R}_1^{n+1}$, since precisely one of the two orthogonal transformations A and $-A$ takes the component H^n to itself. There are two ways to extend an isometry A of H^n to an orthogonal transformation B of $\mathbb{R}_1^{n+2}$ = Span $\{e_1,...,e_{n+2}\}$; namely, one can define Be_2 to be e_2 or $-e_2$. These two extensions induce different Moebius transformations, since they do not differ by a sign. On the other hand, the extension determined by setting $Be_2 = -e_2$ is projectively equivalent to the extension C of $-A$ satisfying $Ce_2 = e_2$. Hence, the group of Moebius transformations induced from isometries of H^n is isomorphic to $O(n,1)$ itself. As before, each of these transformations induces two Lie transformations.

For both the spherical and hyperbolic metrics, there is a parallel transformation P_t which adds t to the signed radius of each sphere while keeping the center fixed. As we saw in Section 1.4, the sphere in S^n with center p and signed radius ρ is represented by the point $[(\cos \rho, p, \sin \rho)]$ in Q^{n+1}. One can easily check that *spherical parallel transformation* P_t is accomplished by the following orthogonal transformation,

$$P_t e_1 = \cos t\, e_1 + \sin t\, e_{n+3} \, ,$$
(5.3) $$P_t e_{n+3} = -\sin t\, e_1 + \cos t\, e_{n+3} \, ,$$
$$P_t e_i = e_i \, , \quad 2 \le i \le n+2 \, .$$

In H^n, the sphere with center $p \in H^n$ and signed radius ρ corresponds to the point $[(p + \cosh \rho\, e_2 + \sinh \rho\, e_{n+3})]$ in Q^{n+1}. *Hyperbolic parallel transformation* P_t is accomplished by the following transformation,

$$P_t e_i = e_i \, , \quad i = 1,3,...,n+2,$$
(5.4) $$P_t e_2 = \cosh t\, e_2 + \sinh t\, e_{n+3} \, ,$$
$$P_t e_{n+3} = \sinh t\, e_2 + \cosh t\, e_{n+3} \, .$$

The following theorem demonstrates the important role played by parallel transformations in generating the Lie sphere group. This result is crucial in the proof that tautness is invariant under Lie transformations (see Section 3.6), and it was first proven in Cecil–Chern [1].

Theorem 5.3: *Any Lie transformation α can be written as $\alpha = \Phi P_t \Psi$, where Φ and Ψ are Moebius transformations and P_t is some Euclidean, spherical or hyperbolic parallel transformation.*

Proof: Represent α by a transformation $A \in O(n+1,2)$. If $Ae_{n+3} = \pm\, e_{n+3}$, then α is a Moebius transformation. If not, then Ae_{n+3} is some unit timelike vector v linearly independent from e_{n+3}. The plane $[e_{n+3}, v]$ in $\mathbb{R}_2^{n+3}$ can have signature $(-, -)$, $(-, +)$ or $(-, 0)$. In the case where the plane has signature $(-,-)$, we can write

$$v = -\sin t\, u_1 + \cos t\, e_{n+3} \, ,$$

where u_1 is a unit timelike vector orthogonal to e_{n+3}, and $t \in (0, \pi)$. Let Φ be a Moebius transformation such that $\Phi^{-1} u_1 = e_1$. Then from equation (5.3), we see that $P_{-t} \Phi^{-1} v = e_{n+3}$. Hence, $P_{-t} \Phi^{-1} \alpha\, e_{n+3} = e_{n+3}$, i.e., $P_{-t} \Phi^{-1} \alpha$ is a Moebius transformation Ψ. Thus, $\alpha = \Phi P_t \Psi$, as desired.

The other two cases are similar. If the plane $[e_{n+3}, v]$ has signature

$(-, 0)$, then we can write

$$v = -t\, u_1 + t\, u_2 + e_{n+3}\,,$$

where u_1 and u_2 are unit timelike and spacelike vectors, respectively, orthogonal to e_{n+3} and to each other. If Φ is a Moebius transformation such that $\Phi^{-1} u_1 = e_1$ and $\Phi^{-1} u_2 = e_2$, then $P_{-t}\,\Phi\alpha$ is a Moebius transformation Ψ, where P_t is the Euclidean parallel transformation given in (4.23). As before, we get $\alpha = \Phi P_t \Psi$. Finally, if the plane $[e_{n+3}, v]$ has signature $(-, +)$, then

$$v = \sinh t\, u_2 + \cosh t\, e_{n+3}\,,$$

for a unit spacelike vector u_2 orthogonal to e_{n+3}. Let Φ be a Moebius transformation such that $\Phi^{-1} u_2 = e_2$, and conclude that $\alpha = \Phi P_t \Psi$ for the hyperbolic parallel transformation P_t in (5.4).   □

3

Legendre Submanifolds

In this chapter, we develop the framework necessary to study submanifolds within the context of Lie sphere geometry. The manifold Λ^{2n-1} of projective lines on the Lie quadric Q^{n+1} has a contact structure, i.e., a globally defined 1–form ω such that $\omega \wedge (d\omega)^{n-1} \neq 0$ on Λ^{2n-1}. This gives rise to a codimension one distribution D on Λ^{2n-1} which has integral submanifolds of dimension $n-1$, but none of higher dimension. These integral submanifolds are called Legendre submanifolds. Any submanifold of a real space–form $\mathbb{R}^n$, S^n or H^n naturally induces a Legendre submanifold, and thus Lie sphere geometry can be used to analyze submanifolds of these spaces. This has been particularly effective in the classification of Dupin submanifolds, which are defined in Section 3.4. In Section 3.5, we define the Lie curvatures of a Legendre submanifold. These are natural Lie invariants which have proven to be valuable in the study of Dupin submanifolds but are defined on the larger class of Legendre submanifolds. We then give a Lie geometric characterization of those Legendre submanifolds which are Lie equivalent to an isoparametric hypersurface in a sphere (Theorem 5.6). In Section 3.6, we prove that tautness is Lie invariant. Finally, in Section 3.7, we discuss the counterexamples of Pinkall–Thorbergsson and Miyaoka–Ozawa to the conjecture that a compact proper Dupin hypersurface in a sphere must be Lie equivalent to an isoparametric hypersurface.

3.1 Contact Structure

In this section, we demonstrate explicitly that the manifold Λ^{2n-1} of projective lines on the Lie quadric Q^{n+1} is a contact manifold. The reader is referred to

Arnold [1, p.349] or Blair [1] for a more complete treatment of contact manifolds in general.

As in earlier chapters, let $e_1,...,e_{n+3}$ denote the standard orthonormal basis for $\mathbb{R}_2^{n+3}$ with e_1 and e_{n+3} timelike. We consider S^n to be the unit sphere in the Euclidean space $\mathbb{R}^{n+1}$ spanned by $e_2,...,e_{n+2}$. A *contact element* on S^n consists of a pair (x, ξ) where $x \in S^n$ and ξ is a unit tangent vector to S^n at x. Thus, the space of contact elements is the unit tangent bundle $T_1 S^n$. We consider $T_1 S^n$ to be the $(2n-1)$–dimensional submanifold of $S^n \times S^n \subset \mathbb{R}^{n+1} \times \mathbb{R}^{n+1}$ given by

$$(1.1) \qquad T_1 S^n = \{(x, \xi) \mid |x| = 1,\ |\xi| = 1,\ x \cdot \xi = 0\}.$$

In general, a $(2n-1)$–dimensional manifold V^{2n-1} is said to be a *contact manifold* if it carries a global 1–form ω such that

$$(1.2) \qquad \omega \wedge (d\omega)^{n-1} \neq 0$$

at all points of V^{2n-1}. The form ω is called a *contact form*, and it is only determined up to multiplication by a non–vanishing smooth function. It is known (see, for example, Blair [1, p.10]) that the unit tangent bundle $T_1 M$ to any n–dimensional Riemannian manifold M is a $(2n-1)$–dimensional contact manifold. The contact form ω defines a codimension one distribution D on V^{2n-1},

$$(1.3) \qquad D_p = \{Y \in T_p V^{2n-1} \mid \omega(Y) = 0\},$$

for $p \in V^{2n-1}$, called the *contact distribution*. This distribution is as far from being integrable as possible, in that there exist integral submanifolds of D of dimension $n-1$ but none of higher dimension (see Theorem 2.1 below).

In our particular case, a tangent vector to $T_1 S^n$ at a point (x, ξ) can be written in the form (X, Z) where

$$(1.4) \qquad X \cdot x = 0,\qquad Z \cdot \xi = 0.$$

Differentiation of the condition $x \cdot \xi = 0$ implies that (X, Z) must also satisfy

(1.5) $$X \cdot \xi + Z \cdot x = 0 \,.$$

We will show that the form ω defined by,

(1.6) $$\omega \, (X, Z) = X \cdot \xi \,,$$

is a contact form on $T_1 S^n$. Thus, at a point (x, ξ), the distribution D is the $(2n-2)$–dimensional space of vectors (X, Z) satisfying $X \cdot \xi = 0$, as well as the equations (1.4) and (1.5). Of course, the equation $X \cdot \xi = 0$ together with (1.5) implies that

(1.7) $$Z \cdot x = 0 \,,$$

for vectors (X, Z) in D. To see that ω satisfies (1.2), we will identify $T_1 S^n$ with the manifold Λ^{2n-1} of lines on Q^{n+1} and compute $d\omega$ using the method of moving frames. The results obtained in this calculation will turn out to be useful in our general study of submanifolds.

   We establish a bijective correspondence between the points of $T_1 S^n$ and the lines on the Q^{n+1} by the map

(1.8) $$(x, \xi) \rightarrow [Y_1(x, \xi), Y_{n+3}(x, \xi)] \,,$$

where

(1.9) $$Y_1(x, \xi) = (1, x, 0) \quad \text{and} \quad Y_{n+3}(x, \xi) = (0, \xi, 1) \,.$$

The points on a line on Q^{n+1} correspond to a parabolic pencil of spheres in S^n. By formula (4.4) of Chapter 1, the point $[Y_1(x, \xi)]$ corresponds to the unique point sphere in the pencil determined by (1.8), and $[Y_{n+3}(x, \xi)]$ corresponds to the unique great sphere in the pencil. Since every line on the quadric contains exactly one point sphere and one great sphere by Corollary 5.5 of Chapter 1, the correspondence (1.8) is bijective. We use the correspondence (1.8) to place a differentiable structure on Λ^{2n-1} in such a way that (1.8) becomes a diffeomorphism.

We now introduce the method of moving frames in the context of Lie sphere geometry, as in Cecil–Chern [1]. The reader is referred to Cartan [1], Griffiths [1], Jensen [1] or Spivak [1, Vol.2, Chap. 7] for an exposition of the general method. Since we want to define frames on the manifold Λ^{2n-1}, it is better to use frames for which some of the vectors are lightlike, rather than orthonormal frames. To facilitate the exposition, we agree on the following range of indices in this section,

(1.10) $1 \leq a, b, c \leq n+3$, $3 \leq i, j, k \leq n+1$.

A *Lie frame* is an ordered set of vectors $Y_1, ..., Y_{n+3}$ in $\mathbb{R}_2^{n+3}$ satisfying the relations,

(1.11) $< Y_a, Y_b > = g_{ab}$,

for

(1.12) $[g_{ab}] = \begin{bmatrix} J & 0 & 0 \\ 0 & I_{n-1} & 0 \\ 0 & 0 & J \end{bmatrix}$,

where I_{n-1} is the identity $(n-1) \times (n-1)$ matrix and

(1.13) $J = \begin{bmatrix} 0 & 1 \\ 1 & 0 \end{bmatrix}$.

If $(y_1, ..., y_{n+3})$ are homogeneous coordinates on $\mathbb{P}^{n+2}$ with respect to a Lie frame, then the Lie metric has the form

(1.14) $< y, y > = 2 (y_1 y_2 + y_{n+2} y_{n+3}) + y_3^2 + ... + y_{n+1}^2$.

The space of all Lie frames can be identified with the group $O(n+1,2)$ of which the Lie sphere group, being isomorphic to $O(n+1,2)/\{\pm I\}$, is a quotient group. In this space, we introduce the *Maurer–Cartan forms* ω_a^b by the equation

$$(1.15) \qquad dY_a = \Sigma \omega_a^b \, Y_b \, ,$$

and we adopt the convention that the sum is always over the repeated index. Differentiating (1.11), we get

$$(1.16) \qquad \omega_{ab} + \omega_{ba} = 0 \, ,$$

where

$$(1.17) \qquad \omega_{ab} = \Sigma g_{bc} \, \omega_a^c \, .$$

Equation (1.16) says that the following matrix is skew–symmetric,

$$(1.18) \qquad [\, \omega_{ab} \,] = \begin{bmatrix} \omega_1^2 & \omega_1^1 & \omega_1^i & \omega_1^{n+3} & \omega_1^{n+2} \\ \omega_2^2 & \omega_2^1 & \omega_2^i & \omega_2^{n+3} & \omega_2^{n+2} \\ \omega_j^2 & \omega_j^1 & \omega_j^i & \omega_j^{n+3} & \omega_j^{n+2} \\ \omega_{n+2}^2 & \omega_{n+2}^1 & \omega_{n+2}^i & \omega_{n+2}^{n+3} & \omega_{n+2}^{n+2} \\ \omega_{n+3}^2 & \omega_{n+3}^1 & \omega_{n+3}^i & \omega_{n+3}^{n+3} & \omega_{n+3}^{n+2} \end{bmatrix} .$$

Taking the exterior derivative of (1.15) yields the *Maurer–Cartan equations,*

$$(1.19) \qquad d\omega_a^b = \Sigma \omega_a^c \wedge \omega_c^b \, .$$

We now produce a contact form on $T_1 S^n$ in the context of moving frames. We want to choose a local frame $Y_1, \dots, Y_{n+3}$ on $T_1 S^n$ with Y_1 and Y_{n+3} given by (1.9). When we transfer this frame to Λ^{2n-1}, it will have the property that for each point $\lambda \in \Lambda^{2n-1}$, the line $[Y_1, Y_{n+3}]$ of the frame at λ is the line on the quadric Q^{n+1} corresponding to λ.

On a sufficiently small open subset U in $T_1 S^n$, we can find smooth mappings,

$$v_i : U \to \mathbb{R}^{n+1}, \quad 3 \le i \le n+1 \, ,$$

such that at each point $(x, \xi) \in U$, the vectors $v_3(x, \xi),...,v_{n+1}(x, \xi)$ are unit vectors orthogonal to each other and to x and ξ. By equations (1.4) and (1.5), we see that the vectors

$$(1.20) \qquad \{(v_i, 0), (0, v_i), (\xi, -x)\}, \quad 3 \le i \le n+1 ,$$

form a basis to the tangent space to $T_1 S^n$ at (x, ξ). We now define a Lie frame on U as follows:

$$
(1.21) \quad
\begin{aligned}
& Y_1 (x, \xi) = (1, x, 0) , \quad Y_{n+3} (x, \xi) = (0, \xi, 1) , \\
& Y_2 (x, \xi) = (-1/2, x/2, 0) , \quad Y_{n+2} (x, \xi) = (0, \xi/2, -1/2) , \\
& Y_i (x, \xi) = (0, v_i (x, \xi), 0) , \quad 3 \le i \le n+1 .
\end{aligned}
$$

We want to determine certain of the Maurer–Cartan forms ω_a^b by computing dY_a on the basis (1.20). In particular, we compute the derivatives dY_1 and dY_{n+3} and find

$$
(1.22) \quad
\begin{aligned}
& dY_1(v_i, 0) = (0, v_i, 0) = Y_i , \\
& dY_1(0, v_i) = (0, 0, 0) , \\
& dY_1(\xi, -x) = (0, \xi, 0) = Y_{n+2} + (1/2) Y_{n+3} ,
\end{aligned}
$$

and

$$
(1.23) \quad
\begin{aligned}
& dY_{n+3}(v_i, 0) = (0, 0, 0) , \\
& dY_{n+3}(0, v_i) = (0, v_i, 0) = Y_i , \\
& dY_{n+3}(\xi, -x) = (0, -x, 0) = (-1/2) Y_1 - Y_2 .
\end{aligned}
$$

Comparing these equations with the equation

$$dY_a = \sum \omega_a^b Y_b ,$$

we see that the 1–forms,

$$(1.24) \qquad \{ \omega_1^i , \omega_{n+3}^i , \omega_1^{n+2} \} ,$$

form the dual basis to the basis given in (1.20) for the tangent space to $T_1 S^n$ at (x, ξ). Since $(\xi, -x)$ has length $\sqrt{2}$, we have

$$\omega_1^{n+2} (X, Z) = ((X, Z) \cdot (\xi, -x)) / 2 = (X \cdot \xi - Z \cdot x) / 2 ,$$

for a tangent vector (X, Z) to $T_1 S^n$ at (x, ξ). Using equation (1.5),

$$X \cdot \xi + Z \cdot x = 0 ,$$

we see that

(1.25)
$$\omega_1^{n+2} (X, Z) = X \cdot \xi ,$$

so ω_1^{n+2} is precisely the form ω in (1.6). We now want to show that ω_1^{n+2} satisfies condition (1.2). This is a straightforward calculation using the Maurer–Cartan equation (1.19) for $d\omega_1^{n+2}$ and the skew–symmetry relations (1.18). By (1.19), we have

$$d\omega_1^{n+2} = \sum \omega_1^c \wedge \omega_c^{n+2} .$$

The skew–symmetry relations imply that $\omega_1^2 = 0$ and $\omega_{n+3}^{n+2} = 0$. Furthermore, in computing $(d\omega_1^{n+2})^{n-1}$, we can ignore any term involving ω_1^{n+2}, since we will eventually take the wedge product with ω_1^{n+2}. Thus, in computing the wedge product $d\omega_1^{n+2} \wedge d\omega_1^{n+2}$, we only need to consider

$$(\sum d\omega_1^i \wedge d\omega_i^{n+2}) \wedge (\sum d\omega_1^j \wedge d\omega_j^{n+2}) .$$

If $i \neq j$, we have a term of the form

$$\omega_1^i \wedge \omega_i^{n+2} \wedge \omega_1^j \wedge \omega_j^{n+2} = \omega_1^i \wedge (- \omega_{n+3}^i) \wedge \omega_1^j \wedge (- \omega_{n+3}^j) = \omega_1^i \wedge \omega_{n+3}^i \wedge \omega_1^j \wedge \omega_{n+3}^j \neq 0 ,$$

where the sign changes are due to (1.18). The last term is non–zero since each of the factors is in the basis (1.24). Thus, we have

$$(1.26) \qquad d\omega_1^{n+2} \wedge d\omega_1^{n+2} = 2 \sum_{i<j} \omega_1^i \wedge \omega_{n+3}^i \wedge \omega_1^j \wedge \omega_{n+3}^j \pmod{\omega_1^{n+2}} .$$

One continues this process by taking the wedge product of (1.26) with $d\omega_1^{n+2}$. This time there are three sign changes in each term as a result of the skew–symmetry relations, and we get

$$(d\omega_1^{n+2})^3 = (-1)^3 (3!) \sum_{i<j<k} \omega_1^i \wedge \omega_{n+3}^i \wedge \omega_1^j \wedge \omega_{n+3}^j \wedge \omega_1^k \wedge \omega_{n+3}^k \pmod{\omega_1^{n+2}} .$$

Continuing this process, one eventually obtains

$$\omega_1^{n+2} \wedge (d\omega_1^{n+2})^{n-1} = \omega_1^{n+2} \wedge (\sum \omega_1^i \wedge \omega_i^{n+2})^{n-1}$$
$$= (-1)^{n-1} (n-1)! \, \omega_1^{n+2} \wedge \omega_1^3 \wedge \omega_{n+3}^3 \wedge \dots \wedge \omega_1^{n+1} \wedge \omega_{n+3}^{n+1} \neq 0 .$$

The last form is non–zero because the set (1.24) is a basis for the cotangent space to $T_1 S^n$ at (x, ξ). Finally, note that the form

$$(1.27) \qquad\qquad \omega_1^{n+2} = \langle dY_1 , Y_{n+3} \rangle ,$$

is globally defined on $T_1 S^n$, since Y_1 and Y_{n+3} are globally defined by (1.21), even though the rest of the Lie frame is only defined on the open set U.

As we noted above, we can use the diffeomorphism (1.8) to transfer this Lie frame and contact form ω_1^{n+2} to the manifold Λ^{2n-1} of lines on the Lie quadric. Now, suppose that $\{Z_1,\dots,Z_{n+3}\}$ is an arbitrary Lie frame on the open set U with the property that the line $[Z_1, Z_{n+3}]$ equals the line $[Y_1, Y_{n+3}]$ at all points of U, i.e.,

$$(1.28) \qquad Z_1 = \alpha Y_1 + \beta Y_{n+3} , \qquad Z_{n+3} = \gamma Y_1 + \delta Y_{n+3} ,$$

for smooth functions α, β, γ, δ with $\alpha\delta - \beta\gamma \neq 0$ on U. Let $\{\theta_a^b\}$ be the Maurer–Cartan forms for this Lie frame. Then, using the scalar product relations (1.11), we get

$$\theta_1^{n+2} = \; < dZ_1 \, , Z_{n+3} > \; = \; < d(\alpha Y_1 + \beta Y_{n+3}) \, , \gamma Y_1 + \delta Y_{n+3} >$$

$$= \alpha\delta < dY_1 \, , Y_{n+3} > + \; \beta\gamma < dY_{n+3} \, , Y_1 > \; = \alpha\delta \, \omega_1^{n+2} + \beta\gamma \, \omega_{n+3}^{2}$$

$$= (\alpha\delta - \beta\gamma) \, \omega_1^{n+2} .$$

Thus, θ_1^{n+2} is also a contact form on $T_1 S^n$.

3.2 Definition of Legendre Submanifolds

In the last section, we showed that $T_1 S^n$ (and hence Λ^{2n-1}) is a contact manifold. We begin this section by proving a basic result concerning contact manifolds in general (Theorem 2.1). Let V^{2n-1} be a contact manifold with contact form ω. Let D be the corresponding contact distribution defined by

$$D_p = \{ \; Y \in T_p V^{2n-1} \; | \; \omega(Y) = 0 \; \} \, ,$$

for $p \in V^{2n-1}$. An immersion $\varphi : W^k \to V^{2n-1}$ of a smooth k–dimensional manifold W^k into V^{2n-1} is called an *integral submanifold* of the distribution D if $\varphi^* \omega = 0$ on W^k, i.e., for each tangent vector Y at each point $w \in W^k$, the vector $d\varphi \, (Y)$ is in the distribution D at the point $\varphi(w)$. (See Blair [1, p.36].)

Theorem 2.1: *Let* V^{2n-1} *be a contact manifold with contact form* ω. *Then there exist integral submanifolds of the contact distribution* D *of dimension* $n-1$, *but none of higher dimension.*

Proof: This is a local result, and the key tool in the local study of contact manifolds is Darboux's Theorem (see, for example, Arnold [1, p.362] or Sternberg [1, p.141]), which states that for every point of a contact manifold V^{2n-1}, there is a local coordinate neighborhood U with coordinates (x_i , y_i , z), $1 \le i \le n-1$, on which the contact form satisfies

$$\omega = dz - \sum_{i=1}^{n-1} y_i \, dx_i .$$

Now, given any point in U with local coordinates (x_i^0 , y_i^0 , z^0), the slice defined by the equations,

$$x_i = \overset{0}{x_i}, \quad z = \overset{0}{z}, \quad 1 \le i \le n-1,$$

clearly defines an $(n-1)$–dimensional integral submanifold of D containing the given point.

Conversely, suppose that W^k is an immersed k–dimensional integral submanifold of D with $k > n-1$. Since this is a local result, we may consider W^k to be embedded in V^{2n-1}. Let $X_1,...,X_k$ be linearly independent vector fields tangent to W^k on some open set Ω in W^k. Let $X_{k+1},...,X_{2n-1}$ be tangent vectors to V^{2n-1} at a point $w \in \Omega$ such that $X_1,...,X_{2n-1}$ form a basis for $T_w V^{2n-1}$. Since the tangent distribution to W^k is integrable, the Lie bracket $[X_i, X_j]$ is also tangent to W^k, for $1 \le i, j \le k$. Furthermore, since the tangent distribution to W^k is contained in the distribution D, we have

$$\omega(X_i) = 0, \quad \omega([X_i, X_j]) = 0,$$

and thus,

$$d\omega(X_i, X_j) = (X_i \omega(X_j) - X_j \omega(X_i) - \omega([X_i, X_j])) = 0,$$

for $1 \le i, j \le k$, on Ω. Since $k > n-1$, this implies that at the point w, we have

$$\omega \wedge (d\omega)^{n-1}(X_1,...,X_{2n-1}) = 0,$$

contradicting the assumption that ω is a contact form on V^{2n-1}.   □

An immersed $(n-1)$–dimensional integral submanifold of the contact distribution D is called a *Legendre submanifold*. We now return to our specific case of the contact manifold $T_1 S^n$. We first want to formulate necessary and sufficient conditions for a smooth map $\mu : M^{n-1} \to T_1 S^n$ to be a Legendre submanifold. We consider $T_1 S^n$ as a submanifold of $S^n \times S^n$ as in (1.1). Thus, we can write $\mu = (f, \xi)$, where f and ξ are both smooth maps from M^{n-1} to S^n.

Theorem 2.2: *A smooth map* $\mu = (f, \xi)$ *from an* $(n-1)$*–dimensional manifold* M^{n-1} *into* $T_1 S^n$ *is a Legendre submanifold if and only if the following three conditions are satisfied.*

(1) *Scalar product conditions:* $f \cdot f = 1$, $\xi \cdot \xi = 1$, $f \cdot \xi = 0$.

(2) *Immersion condition: There is no non-zero tangent vector* X *at any point* $x \in M^{n-1}$ *such that* $df(X)$ *and* $d\xi(X)$ *are both equal to zero.*

(3) *Contact condition:* $df \cdot \xi = 0$.

Proof: By (1.1), the scalar product conditions are precisely the conditions necessary for the image of the map $\mu = (f, \xi)$ to be contained in $T_1 S^n$. Next, since

$$d\mu(X) = (df(X), d\xi(X)),$$

the second condition is precisely what is required for μ to be an immersion. Finally, from (1.6) we have

$$\omega(d\mu(X)) = df(X) \cdot \xi(x),$$

for each $X \in T_x M^{n-1}$. Hence, the condition $\mu^* \omega = 0$ on M^{n-1} is equivalent to the third condition above. □

We now want to translate these conditions to the projective setting, and find necessary and sufficient conditions for a smooth map $\lambda : M^{n-1} \to \Lambda^{2n-1}$ to be a Legendre submanifold. We again make use of the diffeomorphism (1.8) between $T_1 S^n$ and Λ^{2n-1}. For each $x \in M^{n-1}$, $\lambda(x)$ is a line on Q^{n+1}. This line contains exactly one point $[Y_1(x)]$ corresponding to a point sphere in S^n and one point $[Y_{n+3}(x)]$ corresponding to a great sphere in S^n. The map $[Y_1]$ from M^{n-1} to Q^{n+1} is called the *Moebius projection* or *point sphere map* of λ, and likewise, $[Y_{n+3}]$ is called the *great sphere map*. The homogeneous coordinates of these points with respect to the standard basis are given by

(2.1) $\qquad Y_1(x) = (1, f(x), 0)$, $\quad Y_{n+3}(x) = (0, \xi(x), 1)$, where f and ξ are smooth maps from M^{n-1} defined by formula (2.1). The map f is called the *spherical projection* of λ, and ξ is called the *spherical field of unit normals*. The maps f and ξ depend on the choice of orthonormal basis $\{e_1,...,e_{n+2}\}$ for the orthogonal complement of e_{n+3} . In this way, λ determines a map $\mu = (f, \xi)$ from M^{n-1} to $T_1 S^n$, and because of the diffeomorphism (1.8), λ is a Legendre submanifold if and only if μ satisfies the conditions of Theorem 2.2. However, it is useful to have conditions for when λ determines a Legendre submanifold which do not depend on the special parametrization of λ by $\{Y_1, Y_{n+3}\}$. In fact, in most applications of Lie geometry to submanifolds of S^n or $\mathbb{R}^n$, it is better to use a Lie frame $Z_1,...,Z_{n+3}$ with $\lambda = [Z_1, Z_{n+3}]$, where Z_1 and Z_{n+3} are not the point sphere and great sphere maps. The following projective formulation of the conditions needed for a Legendre submanifold was given by Pinkall [4], where he referred to a Legendre submanifold as a "Lie geometric hypersurface." The three conditions correspond, respectively, to the three conditions in Theorem 2.2.

Theorem 2.3: *Let* $\lambda : M^{n-1} \to \Lambda^{2n-1}$ *be a smooth map with* $\lambda = [Z_1, Z_{n+3}]$, *where* Z_1 *and* Z_{n+3} *are smooth maps from* M^{n-1} *into* $\mathbb{R}_2^{n+3}$. *Then,* λ *is a Legendre submanifold if and only if* Z_1 *and* Z_{n+3} *satisfy the following conditions:*

(1) *Scalar product conditions: For each* $x \in M^{n-1}$, *the vectors* $Z_1(x)$ *and* $Z_{n+3}(x)$ *are linearly independent and* $< Z_i , Z_j > = 0$, *for* $i, j \in \{1, n+3\}$.

(2) *Immersion condition: There is no non-zero tangent vector* X *at any point* $x \in M^{n-1}$ *such that* $dZ_1(X)$ *and* $dZ_{n+3}(X) \in Span \{Z_1(x), Z_{n+3}(x)\}$.

(3) *Contact condition:* $< dZ_1 , Z_{n+3} > = 0$.

These conditions are invariant under a reparametrization $\lambda = [W_1, W_{n+3}]$, *where* $W_1 = \alpha Z_1 + \beta Z_{n+3}$, $W_{n+3} = \gamma Z_1 + \delta Z_{n+3}$, *for smooth functions* α, β, γ, δ *on* M^{n-1} *with* $\alpha\delta - \beta\gamma \neq 0$.

Proof: The proof is accomplished in two steps. First, we know that the map λ induces two maps Y_1 and Y_{n+3} as in (2.1), which in turn determine f and ξ. We show that the pair $\{Y_1, Y_{n+3}\}$ satisfies (1) – (3) if and only if the map $\mu = (f, \xi)$ satisfies the conditions in Theorem 2.2. Secondly, we show that the conditions (1) – (3) are invariant under a transformation to $\{W_1, W_{n+3}\}$ as above. In particular, the conditions are satisfied by an arbitrary pair $\{Z_1, Z_{n+3}\}$ if and only if they are satisfied by $\{Y_1, Y_{n+3}\}$.

First, consider $Y_1 = (1, f, 0)$, $Y_{n+3} = (0, \xi, 1)$. Then,

$$< Y_1, Y_1 > = |f|^2 - 1, \quad < Y_{n+3}, Y_{n+3} > = |\xi|^2 - 1, \quad < Y_1, Y_{n+3} > = f \cdot \xi .$$

Thus, condition (1) for the pair $\{Y_1, Y_{n+3}\}$ is equivalent to the scalar product condition for the pair (f, ξ) in Theorem 2.2. Next, suppose that X is a non-zero vector in $T_x M^{n-1}$. Then,

$$(2.2) \qquad dY_1(X) = (0, df(X), 0) , \quad dY_{n+3}(X) = (0, d\xi(X), 0) .$$

Hence, $dY_1(X)$ and $dY_{n+3}(X)$ are in Span $\{Y_1(x), Y_{n+3}(x)\}$ if and only if they are zero, i.e., $df(X) = 0$ and $d\xi(X) = 0$. Therefore, the pair $\{Y_1, Y_{n+3}\}$ satisfies (2) if and only if the pair (f, ξ) satisfies the immersion condition of Theorem 2.2. Finally, from (2.2), we have

$$< dY_1(X), Y_{n+3}(x) > = df(X) \cdot \xi(x) ,$$

so $\{Y_1, Y_{n+3}\}$ satisfies (3) if and only if (f, ξ) satisfies the contact condition of Theorem 2.2.

Now, suppose that a pair $\{Z_1, Z_{n+3}\}$ satisfies conditions (1) – (3) and that $\{W_1, W_{n+3}\}$ is given as in the statement of the theorem. It follows from Theorem 5.4 of Chapter 1 that (1) is precisely the condition that $[Z_1(x)]$ and $[Z_{n+3}(x)]$ be two distinct points in Q^{n+1} such that the line $[Z_1(x), Z_{n+3}(x)]$ lies on Q^{n+1}. This clearly holds for the pair $\{W_1, W_{n+3}\}$ if and only if it holds for $\{Z_1, Z_{n+3}\}$, since $[W_1, W_{n+3}] = [Z_1, Z_{n+3}]$. Next, we compute

$$dW_1 = \alpha \, dZ_1 + \beta \, dZ_{n+3} + (d\alpha) \, Z_1 + (d\beta) \, Z_{n+3} ,$$

(2.3)

$$dW_{n+3} = \gamma \, dZ_1 + \delta \, dZ_{n+3} + (d\gamma) \, Z_1 + (d\delta) \, Z_{n+3} .$$

The condition $\alpha\delta - \beta\gamma \neq 0$ allows us to solve for dZ_1 and dZ_{n+3} in terms of dW_1 and dW_{n+3} mod $\{Z_1 , Z_{n+3}\}$. From this, it is clear that (2) holds for the pair $\{W_1, W_2\}$ if and only if it holds for $\{Z_1, Z_{n+3}\}$. Finally, note that the scalar product condition $< Z_1, Z_{n+3} > = 0$ implies that

$$< dZ_{n+3}, Z_1 > = - < dZ_1, Z_{n+3} > .$$

Using this and the scalar product relations (1), we compute

$$
\begin{aligned}
< dW_1, W_{n+3} > &= < \alpha \, dZ_1 + \beta \, dZ_{n+3} + (d\alpha) \, Z_1 + (d\beta) \, Z_{n+3} \, , \gamma \, Z_1 + \delta \, Z_{n+3} > \\
&= \alpha\delta < dZ_1 , Z_{n+3} > + \beta\gamma < dZ_{n+3} , Z_1 > \\
&= (\alpha\delta - \beta\gamma) < dZ_1 , Z_{n+3} > .
\end{aligned}
$$

Thus, $\{W_1, W_{n+3}\}$ satisfies (3) precisely when $\{Z_1, Z_{n+3}\}$ satisfies (3).   □

3.3 The Legendre Map

All oriented hypersurfaces in the sphere S^n, Euclidean space $\mathbb{R}^n$, and hyperbolic space H^n naturally induce Legendre submanifolds of Λ^{2n-1}, as do all submanifolds of codimension $m > 1$ in these spaces. In this section, we study these examples and see, conversely, how a Legendre submanifold naturally induces a smooth map into S^n which may have singularities.

First, suppose that $f : M^{n-1} \to S^n$ is an immersed oriented hypersurface with field of unit normals $\xi : M^{n-1} \to S^n$. The induced Legendre submanifold is given by the map $\lambda : M^{n-1} \to \Lambda^{2n-1}$ defined by $\lambda(x) = [Y_1(x), Y_{n+3}(x)]$, where

(3.1) $Y_1(x) = (1, f(x), 0) , \qquad Y_{n+3}(x) = (0, \xi(x), 1) .$

The map λ is called the *Legendre map* induced by the immersion f with field of unit normals ξ. It is easy to check that the pair $\{Y_1, Y_{n+3}\}$ satisfies the conditions of Theorem 2.3. Condition (1) is immediate since both f and ξ are maps into S^n, and $\xi(x)$ is tangent to S^n at $f(x)$ for each x in M^{n-1}. Condition (2) is satisfied since

$$dY_1(X) = (0, df(X), 0) ,$$

for any vector $X \in T_x M^{n-1}$. Since f is an immersion, $df(X) \neq 0$ for a non–zero vector X, and thus $dY_1(X)$ is not in Span $\{Y_1(x), Y_{n+3}(x)\}$. Finally, condition (3) is satisfied since

$$< dY_1(X), Y_{n+3}(x) > = df(X) \cdot \xi(x) = 0 ,$$

because ξ is a field of normals to f.

Next, we handle the case of a submanifold $\varphi : V \to S^n$ of codimension $m+1$ greater than 1. Let B^{n-1} be the unit normal bundle of the submanifold φ. Then B^{n-1} can be considered to be the submanifold of $V \times S^n$ given by

$$B^{n-1} = \{(x, \xi) \mid \varphi(x) \cdot \xi = 0, \ d\varphi(X) \cdot \xi = 0, \text{ for all } X \in T_x V\} .$$

The induced Legendre submanifold $\lambda : B^{n-1} \to \Lambda^{2n-1}$ is defined by

$$(3.2) \qquad \lambda(x, \xi) = [Y_1(x, \xi), Y_{n+3}(x, \xi)] ,$$

where

$$(3.3) \qquad Y_1(x, \xi) = (1, \varphi(x), 0) , \qquad Y_{n+3}(x, \xi) = (0, \xi, 1) .$$

Geometrically, $\lambda(x, \xi)$ is the line on Q^{n+1} corresponding to the parabolic pencil of spheres in S^n in oriented contact at the contact element $(\varphi(x), \xi) \in T_1 S^n$. As in the case of a hypersurface, condition (1) is easily checked. However, condition (2) is somewhat different. To compute the differentials of Y_1 and Y_{n+3} at a given point (x, ξ), we first construct a local trivialization of B^{n-1} in a neighborhood of (x, ξ). Let $\xi_0,...,\xi_m$ be an orthonormal normal frame at x with $\xi_0 = \xi$. Let W be a normal coordinate neighborhood of x in V, as defined in Kobayashi–Nomizu [1, Vol.1, p.148], and extend $\xi_0,...,\xi_m$ to orthonormal normal vector fields on W by parallel translation with respect to the normal connection along geodesics in V through x. For any point $w \in W$ and unit normal η to $\varphi(V)$ at w, we can write,

$$\eta = \left[1 - \sum_{i=1}^{m} t_i^2 \right]^{1/2} \xi_0 + t_1 \xi_1 + \ldots + t_m \xi_m \ ,$$

where $0 \le |t_i| \le 1$, for all i, and $t_1^2 + \ldots + t_m^2 \le 1$. The tangent space to B^{n-1} at the given point (x, ξ) can be considered to be

(3.4) $T_x V \times \text{Span} \ \{ \ \partial/\partial t_1, \ldots, \partial/\partial t_m \ \} = T_x V \times \mathbb{R}^m$.

Since $\xi_0(x) = \xi$ and ξ_0 is parallel with respect to the normal connection, we have for $X \in T_x V$,

$$d\xi_0(X) = d\varphi \ (- A^{\xi} X) \ ,$$

where A^{ξ} is the shape operator determined by ξ. Thus, we have,

(3.5) $dY_1(X, 0) = (0, d\varphi(X), 0)$,
            $dY_{n+3}(X, 0) = (0, d\xi_0(X), 0) = (0, d\varphi \ (-A^{\xi}X), 0)$.

Next, we compute from (3.3),

(3.6) $dY_1(0, Z) = (0, 0, 0)$, $dY_{n+3}(0, Z) = (0, Z, 0)$.

From (3.5) and (3.6), we see that there is no non−zero vector (X, Z) such that $dY_1(X, Z)$ and $dY_{n+3}(X, Z)$ are both in Span $\{Y_1, Y_{n+3}\}$, and so (2) is satisfied. Finally, condition (3) holds since

$$< dY_1(X, Z), Y_{n+3}(x, \xi) > \ = d\varphi(X) \cdot \xi = 0 \ .$$

The situation for submanifolds of $\mathbb{R}^n$ or H^n is similar. First, suppose that $F:M^{n-1} \to \mathbb{R}^n$ is an oriented hypersurface with field of unit normals $\eta:M^{n-1} \to \mathbb{R}^n$. As usual, we identify $\mathbb{R}^n$ with the subspace of $\mathbb{R}_2^{n+3}$ spanned by $e_3, \ldots, e_{n+2}$. The induced Legendre submanifold $\lambda : M^{n-1} \to \Lambda^{2n-1}$ is defined by $\lambda = [Y_1, Y_{n+3}]$, where

(3.7) $Y_1 = (1 + F \cdot F , 1 - F \cdot F , 2F , 0) / 2 , \quad Y_{n+3} = (F \cdot \eta, - (F \cdot \eta), \eta, 1)$.

By (3.4) of Chapter 1, $[Y_1(x)]$ corresponds to the point sphere and $[Y_{n+3}(x)]$ corresponds to the hyperplane in the parabolic pencil determined by the line $\lambda(x)$, for each $x \in M^{n-1}$. The reader can easily verify conditions (1) – (3) of Theorem 2.3 in a manner similar to the spherical case. In the case of a submanifold $\psi : V \to \mathbb{R}^n$ of codimension greater than one, the induced Legendre submanifold is the map λ from the unit normal bundle B^{n-1} to Λ^{2n-1} defined by $\lambda(x, \eta) = [Y_1(x, \eta), Y_{n+3}(x, \eta)]$, where

$$Y_1(x, \eta) = (1 + \psi(x) \cdot \psi(x) , 1 - \psi(x) \cdot \psi(x) , 2\psi(x) , 0) / 2 ,$$

(3.8)

$$Y_{n+3}(x, \eta) = (\psi(x) \cdot \eta , - (\psi(x) \cdot \eta) , \eta , 1) .$$

The verification that the pair $\{Y_1, Y_{n+3}\}$ satisfies (1) – (3) is similar to that for submanifolds of S^n of codimension greater that one.

Finally, as in Section 1.4, we consider H^n to be the submanifold of the Lorentz space $\mathbb{R}_1^{n+1}$ spanned by $e_1, e_3, ..., e_{n+2}$ defined as follows:

$$H^n = \{ y \in \mathbb{R}_1^{n+1} \mid (y, y) = - 1, \ y_1 \geq 1 \} ,$$

where $(,)$ is the Lorentz metric on $\mathbb{R}_1^{n+1}$ obtained by restricting the Lie metric. Let $h : M^{n-1} \to H^n$ be a hypersurface with field of unit normals $\zeta : M^{n-1} \to \mathbb{R}_1^{n+1}$. The induced Legendre submanifold is given by the map $\lambda : M^{n-1} \to \Lambda^{2n-1}$ defined by $\lambda = [Y_1, Y_{n+3}]$, where

(3.9) $Y_1(x) = h(x) + e_2 , \quad Y_{n+3}(x) = \zeta(x) + e_{n+3}$.

Note that $(h, h) = -1$, so $<Y_1, Y_1> = 0$, while $(\zeta, \zeta) = 1$, so $<Y_{n+3}, Y_{n+3}> = 0$. The reader can easily check that the conditions (1) – (3) are satisfied. Finally, if $\gamma : V \to H^n$ is an immersed submanifold of codimension greater than one, then the induced Legendre submanifold $\lambda : B^{n-1} \to \Lambda^{2n-1}$ is again defined on the unit normal bundle to the submanifold γ in the obvious way.

Now, suppose that $\lambda : M^{n-1} \to \Lambda^{2n-1}$ is an arbitrary Legendre submanifold.

As we have seen, it is always possible to parametrize λ by the point sphere map $[Y_1]$ and the great sphere map $[Y_{n+3}]$ given by

$$(3.10) \qquad Y_1 = (\ 1, f, 0\ )\ , \qquad Y_{n+3} = (\ 0, \xi, 1\ )\ .$$

This defines the two maps f and ξ from M^{n-1} to S^n, which we called the spherical projection and spherical field of unit normals, respectively, in Section 3.2. Both f and ξ are smooth maps, but neither need be an immersion or even have constant rank. (See Example 3.1 below.) The Legendre submanifold induced from an oriented hypersurface in S^n is the special case where the spherical projection f is an immersion, i.e., has constant rank $n-1$ on M^{n-1}. In the case of the Legendre submanifold induced from a submanifold $\varphi:V^k \to S^n$, the spherical projection $f : B^{n-1} \to S^n$ defined by $f(x, \xi) = \varphi(x)$ has constant rank k.

If the range of the locus of point spheres $[Y_1]$ does not contain the improper point $[(1,-1,0,...,0)]$, then λ also determines a *Euclidean projection* $F:M^{n-1} \to \mathbb{R}^n$ and a *Euclidean field of unit normals* $\eta : M^{n-1} \to \mathbb{R}^n$. These are defined by the equation $\lambda = [Z_1, Z_{n+3}]$, where

$$(3.11) \quad Z_1 = (\ 1 + F\cdot F\ ,\ 1 - F\cdot F\ ,\ 2F, 0\ )\ /\ 2\ , \quad Z_{n+3} = (\ F\cdot\eta, -(F\cdot\eta), \eta, 1\ )\ .$$

Here $[Z_1(x)]$ corresponds to the unique point sphere in the parabolic pencil determined by $\lambda(x)$, and $[Z_{n+3}(x)]$ corresponds to the unique plane in this pencil. As in the spherical case, the smooth maps F and η need not have constant rank. Finally, if the range of the Euclidean projection F lies inside some disk Ω in $\mathbb{R}^n$, then one can define a hyperbolic projection and hyperbolic field of unit normals by placing a hyperbolic metric on Ω.

Example 3.1: *A Euclidean projection which is not an immersion.*

An example where the Euclidean (or spherical) projection does not have constant rank is illustrated by the cyclide of Dupin in Figure 3.1. Here, the

Legendre submanifold is a map $\lambda : T^2 \to \Lambda^5$, where T^2 is a 2–dimensional torus. The Euclidean projection $F : T^2 \to \mathbb{R}^3$ maps the circle S^1, containing the points A, B, C and D to the point P. But the map λ into the space of contact elements is an immersion. The four arrows in Figure 3.1 represent the contact elements corresponding under the map λ to the four points indicated on the circle S^1. Actually, examples of Legendre submanifolds whose Euclidean or spherical projection is not an immersion are plentiful, as will be seen in the next section.

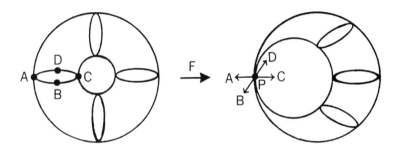

Figure 3.1 – A Euclidean projection F with a singularity

3.4 Curvature Spheres and Parallel Submanifolds

To motivate the definition of a curvature sphere, we consider the case of an oriented hypersurface $f : M^{n-1} \to S^n$ with field of unit normals $\xi : M^{n-1} \to S^n$. The shape operator of f at a point $x \in M^{n-1}$ is the symmetric linear transformation $A : T_x M^{n-1} \to T_x M^{n-1}$ defined by the equation,

(4.1) $df(AX) = - d\xi(X)$.

The eigenvalues of A are called the *principal curvatures*, and the corresponding eigenvectors are called *principal vectors*. We next recall the notion of a focal point of the immersion. For each real number t, define a map $f_t : M^{n-1} \to S^n$ by

(4.2) $f_t = \cos t f + \sin t \, \xi$.

For each $x \in M^{n-1}$, the point $f_t(x)$ lies an oriented distance t along the normal geodesic to $f(M^{n-1})$ at $f(x)$. A point $p = f_t(x)$ is called a *focal point* of multiplicity $m > 0$ of f at x if the nullity of df_t is equal to m at x. Geometrically, one thinks of focal points as points where nearby normal geodesics intersect. It is well known that the location of focal points is related to the principal curvatures. Specifically, if $X \in T_x M^{n-1}$, then by (4.1) we have

(4.3) $df_t(X) = \cos t \, df(X) + \sin t \, d\xi(X) = df(\cos t \, X - \sin t \, AX)$.

Thus, $df_t(X)$ equals zero for $X \neq 0$ if and only if $\cot t$ is a principal curvature of f at x, and X is a corresponding principal vector. Hence, $p = f_t(x)$ is a focal point of f at x of multiplicity m if and only if $\cot t$ is a principal curvature of multiplicity m at x. Note that each principal curvature $\kappa = \cot t$, $0 < t < \pi$, produces two distinct antipodal focal points on the normal geodesic, with parameter values t and $t + \pi$. The oriented hypersphere, centered at a focal point p and in oriented contact with $f(M^{n-1})$ at $f(x)$, is called a *curvature sphere* of f at x. The two antipodal focal points determined by κ are the two centers of the corresponding curvature sphere. Thus, the correspondence between principal curvatures and distinct curvature spheres is bijective. The multiplicity of the curvature sphere is, by definition, equal to the multiplicity of the principal curvature.

We now consider these ideas as they apply to the Legendre submanifold λ induced from the oriented hypersurface determined by f and ξ. As in (3.1), we have $\lambda = [Y_1, Y_{n+3}]$, where

$$(4.4) \qquad\qquad Y_1 = (1, f, 0), \qquad Y_{n+3} = (0, \xi, 1).$$

For each $x \in M^{n-1}$, the points on the line $\lambda(x)$ can be parametrized as

$$(4.5) \qquad [K_t(x)] = [\cos t \; Y_1(x) + \sin t \; Y_{n+3}(x)] = [(\cos t, f_t(x), \sin t)],$$

where f_t is given in (4.2). By (4.4) of Chapter 1, the point $[K_t(x)]$ in Q^{n+1} corresponds to the oriented sphere in S^n with center $f_t(x)$ and signed radius t. This sphere is in oriented contact with the oriented hypersurface $f(M^{n-1})$ at $f(x)$. Given a tangent vector $X \in T_x M^{n-1}$, we have

$$(4.6) \qquad\qquad dK_t(X) = (0, df_t(X), 0).$$

Thus, $dK_t(X) = (0, 0, 0)$ if and only if $df_t(X) = 0$, i.e., $p = f_t(x)$ is a focal point of f at x. Hence, we have shown the following.

Lemma 4.1: *The point $[K_t(x)]$ in Q^{n+1} corresponds to a curvature sphere of the hypersurface f at x if and only if $dK_t(X) = (0, 0, 0)$, for some non-zero vector X in $T_x M^{n-1}$.*

This characterization of curvature spheres depends on the special parametrization of λ given by $\{Y_1, Y_{n+3}\}$, and it has only been defined in the case where the spherical projection f is an immersion. Since it is often desirable to use a different parametrization of λ, we would like a definition of curvature sphere which is independent of the parametrization of λ. We would also like a definition which is valid for an arbitrary Legendre submanifold. This definition is given in the following paragraph.

Let $\lambda : M^{n-1} \to \Lambda^{2n-1}$ be a Legendre submanifold parametrized by the pair $\{Z_1, Z_{n+3}\}$, as in Theorem 2.3. Let $x \in M^{n-1}$ and $r, s \in \mathbb{R}$ with $(r, s) \neq (0, 0)$. The sphere,

$$[K] = [r\, Z_1(x) + s\, Z_{n+3}(x)] \,,$$

is called a *curvature sphere* of λ at x, if there exists a non–zero vector X in $T_x M^{n-1}$ such that

(4.7) $r\, dZ_1(X) + s\, dZ_{n+3}(X) \in \text{Span} \{Z_1(x), Z_{n+3}(x)\} \,.$

The vector X is called a *principal vector* corresponding to $[K]$. By (2.3), this definition is invariant under a change of parametrization of the form considered in Theorem 2.3. Furthermore, if we take the special parametrization $Z_1 = Y_1$, $Z_{n+3} = Y_{n+3}$ given in (4.4), then (4.7) holds if and only if $r\, dY_1(X) + s\, dY_{n+3}(X)$ actually equals $(0, 0, 0)$. Thus, this definition is a generalization of the condition in Lemma 4.1.

From (4.7), it is clear that the set of principal vectors corresponding to a given curvature sphere $[K]$ at x is a subspace of $T_x M^{n-1}$. This set is called the *principal space* corresponding to $[K]$. Its dimension is the *multiplicity* of $[K]$.

Remark 4.2:   The definition of curvature sphere can be developed in the context of Lie geometry without any reference to submanifolds of S^n (see Cecil–Chern [1] for details).   In that case, one begins with a Legendre submanifold $\lambda : M^{n-1} \to \Lambda^{2n-1}$ and considers a curve $\gamma(t)$ lying on M^{n-1}. The set of points in Q^{n+1} lying on the set of lines $\lambda(\gamma(t))$ forms a ruled surface in Q^{n+1}. One then considers conditions for this ruled surface to be developable. This leads to a system of linear equations whose roots determine the curvature spheres at each point along the curve.

We next want to show that the notion of curvature sphere is invariant under Lie transformations. Let $\lambda : M^{n-1} \to \Lambda^{2n-1}$ be a Legendre submanifold parametrized by $\lambda = [Z_1, Z_{n+3}]$. Suppose that $\beta = P(B)$ is the Lie sphere transformation induced by an orthogonal transformation $B \in O(n+1,2)$. Since

B is orthogonal, it is easy to check that the maps, $W_1 = BZ_1$, $W_{n+3} = BZ_{n+3}$, satisfy the conditions (1) – (3) of Theorem 2.3. We will denote the Legendre submanifold defined by $\{W_1, W_{n+3}\}$ by $\beta\lambda : M^{n-1} \to \Lambda^{2n-1}$. The Legendre submanifolds λ and $\beta\lambda$ are said to be *Lie equivalent*. In terms of Euclidean geometry, suppose that V and W are two immersed submanifolds of $\mathbb{R}^n$ (or of S^n or H^n). We say that V and W are *Lie equivalent* if their induced Legendre submanifolds are Lie equivalent.

Consider λ and β as above, so that $\lambda = [Z_1, Z_{n+3}]$ and $\beta\lambda = [W_1, W_{n+3}]$. Note that for a tangent vector $X \in T_x M^{n-1}$ and for real numbers $(r, s) \neq (0, 0)$, we have

(4.8) $r\, dW_1(X) + s\, dW_{n+3}(X) = B\, (\, r\, dZ_1(X) + s\, dZ_{n+3}(X)\,)$,

since B is linear. Thus, we see that

$$r\, dW_1(X) + s\, dW_{n+3}(X) \in \text{Span}\, \{W_1(x), W_{n+3}(x)\}$$

if and only if

$$r\, dZ_1(X) + s\, dZ_{n+3}(X) \in \text{Span}\, \{Z_1(x), Z_{n+3}(x)\}\, .$$

This immediately implies the following theorem.

Theorem 4.3: *Let* $\lambda : M^{n-1} \to \Lambda^{2n-1}$ *be a Legendre submanifold and* β *a Lie sphere transformation. The point* $[K]$ *on the line* $\lambda(x)$ *is a curvature sphere of* λ *at* x *if and only if the point* $\beta\,[K]$ *is a curvature sphere of the Legendre submanifold* $\beta\lambda$ *at* x. *Furthermore, the principal spaces corresponding to* $[K]$ *and* $\beta\,[K]$ *are identical.*

An important special case is that of a Lie transformation determined by a spherical parallel transformation P_t, as given in Section 2.5,

(4.9)
$$P_t e_1 = \cos t \, e_1 + \sin t \, e_{n+3} \, ,$$
$$P_t e_{n+3} = - \sin t \, e_1 + \cos t \, e_{n+3} \, ,$$
$$P_t e_i = e_i \, , \quad 2 \le i \le n+2 \, .$$

We will also denote the induced Lie transformation by P_t. Recall that P_t has the effect of adding t to the signed radius of each sphere in S^n, while keeping the center fixed.

Suppose that $\lambda : M^{n-1} \to \Lambda^{2n-1}$ is a Legendre submanifold parametrized by the point sphere and great sphere maps (4.4). Then $P_t \lambda = [W_1, W_{n+3}]$, where

(4.10) $W_1 = P_t Y_1 = (\cos t, f, \sin t) \, , \quad W_{n+3} = P_t Y_{n+3} = (- \sin t, \xi, \cos t) \, .$

Note that W_1 and W_{n+3} are not the point sphere and great sphere maps for $P_t \lambda$. Solving for the point sphere map Z_1 and the great sphere map Z_{n+3} of $P_t \lambda$, we find

(4.11)
$$Z_1 = \cos t \, W_1 - \sin t \, W_{n+3} = (\, 1 \, , \cos t f - \sin t \, \xi \, , 0 \,) \, ,$$

$$Z_{n+3} = \sin t \, W_1 + \cos t \, W_{n+3} = (\, 0 \, , \sin t f + \cos t \, \xi \, , 1 \,) \, .$$

From this, we see that $P_t \lambda$ has spherical projection and spherical unit normal field given, respectively, by

(4.12)
$$f_{-t} = \cos t f - \sin t \, \xi = \cos (-t) f + \sin (-t) \, \xi \, ,$$

$$\xi_{-t} = \sin t f + \cos t \, \xi = - \sin (-t) f + \cos (-t) \, \xi \, .$$

The minus sign occurs because P_t takes a sphere with center $f_{-t}(x)$ and radius $-t$ to the point sphere $f_{-t}(x)$. We call $P_t \lambda$ a *parallel submanifold* of λ. Formula (4.12) shows the close correspondence between these parallel submanifolds and the "parallel hypersurfaces" f_t to f, in the case where f is an immersed hypersurface. The spherical projection f_t will develop singularities at the focal points of f, but the parallel submanifold $P_t \lambda$ will still be a smooth submanifold of Λ^{2n-1}. The following theorem, due to Pinkall [4, p.428], shows that the number of these singularities is bounded for each $x \in M^{n-1}$.

Theorem 4.4: *Let* $\lambda : M^{n-1} \to \Lambda^{2n-1}$ *be a Legendre submanifold with spherical projection* f *and spherical unit normal field* ξ. *Then for each* $x \in M^{n-1}$, *the parallel map,*

$$f_t = \cos t\, f + \sin t\, \xi ,$$

fails to be an immersion at x for at most n–1 values of $t \in [0, \pi)$.

Here, $[0, \pi)$ is the appropriate interval because of the phenomenon mentioned earlier that each principal curvature of an immersion produces two distinct antipodal focal points in the interval $[0, 2\pi)$. Before proving Pinkall's theorem, we state some important consequences which are obtained by passing to a parallel submanifold, if necessary, and then applying well–known results concerning immersed hypersurfaces in S^n.

Corollary 4.5: *Let* $\lambda : M^{n-1} \to \Lambda^{2n-1}$ *be a Legendre submanifold. Then,*

(a) *At each point* $x \in M^{n-1}$, *there are at most n–1 distinct curvature spheres* $K_1,...,K_g$.

(b) *The principal vectors corresponding to a curvature sphere* K_i *form a subspace* T_i *of the tangent space* $T_x M^{n-1}$.

(c) *The tangent space* $T_x M^{n-1} = T_1 \oplus ... \oplus T_g$.

(d) *If the dimension of a given* T_i *is constant on an open subset* U *of* M^{n-1}, *then the distribution* T_i *is integrable on U.*

(e) *If* $\dim T_i = m > 1$ *on an open subset* U *of* M^{n-1}, *then the curvature sphere* K_i *is constant along the leaves of the foliation* T_i.

Proof: In the case where the spherical projection f of λ is an immersion, the corollary follows from known results concerning hypersurfaces in S^n and the correspondence between curvature spheres of λ and principal curvatures of f. Specifically, (a) − (c) follow from elementary linear algebra applied to the (symmetric) shape operator A of the immersion f. As to (d) and (e), Nomizu [2] showed that any principal curvature function κ_i which has constant multiplicity on an open subset U in M^{n-1} is smooth, as is its corresponding principal distribution T_i. If the multiplicity m_i of κ_i on U is one, then T_i is integrable by the theory of ordinary differential equations. If $m_i > 1$, then the integrability of T_i, and the fact that κ_i is constant along the leaves of T_i are consequences of Codazzi's equation (Ryan [1], see also Cecil–Ryan [7, p.139] and Reckziegel [1] – [3]).

Note that (a) – (c) are pointwise statements, while (d) – (e) hold on U if they can be shown to hold in a neighborhood of each point of U. Now, let x be an arbitrary point of M^{n-1}. If the spherical projection f is not an immersion at x, then by Theorem 4.4, we can find a parallel transformation P_{-t} such that the spherical projection f_t of the Legendre submanifold $P_{-t}\lambda$ is an immersion at x, and hence in a neighborhood of x. Thus, the corollary holds for $P_{-t}\lambda$ in this neighborhood of x. By Theorem 4.3, the corollary also holds for λ in this neighborhood of x. Since x is an arbitrary point, the result is proven. $\square$

Let $\lambda : M^{n-1} \to \Lambda^{2n-1}$ be an arbitrary Legendre submanifold. A connected submanifold S of M^{n-1} is called a *curvature surface* if at each $x \in S$, the tangent space $T_x S$ is equal to a principal space T_i. For example, if dim T_i is constant on an open subset U of M^{n-1}, then each leaf of the principal foliation T_i is a curvature surface on U. Curvature surfaces are plentiful, since the results of Reckziegel [2] or Singley [1] imply that there is an open dense subset Ω of M^{n-1} on which the multiplicities of the curvature spheres are locally constant. On Ω, each leaf of each principal foliation is a curvature surface.

It is also possible to have a curvature surface S which is not a leaf of a principal foliation, because the multiplicity of the corresponding curvature sphere is not constant, as in the following example.

Example 4.6: *A curvature surface which is not a leaf of a principal foliation.*

Let T^2 be a torus of revolution in $\mathbb{R}^3$, and embed $\mathbb{R}^3$ into $\mathbb{R}^4 = \mathbb{R}^3 \times \mathbb{R}$. Let η be a field of unit normals to T^2 in $\mathbb{R}^3$. Let M^3 be a tube of sufficiently small radius $\varepsilon > 0$ around T^2 in $\mathbb{R}^4$, so that M^3 is a compact smooth embedded hypersurface in $\mathbb{R}^4$. The normal space to T^2 in $\mathbb{R}^4$ at a point $x \in T^2$ is spanned by $\eta(x)$ and $e_4 = (0, 0, 0, 1)$. The shape operator A^η has two distinct principal curvatures at each point of T^2, while the shape operator for e_4 is identically zero. Thus, the shape operator B for the normal

$$\zeta = \cos \theta \, \eta(x) + \sin \theta \, e_4 \, ,$$

at a point $x \in T^2$, is given by $B = \cos \theta \, A^{\eta(x)}$. From the formulas for the principal curvatures of a tube (see Cecil–Ryan [7, p.131]), one finds that at all points of M^3 where $x_4 \neq \pm \varepsilon$, there are three distinct principal curvatures, which are constant along their corresponding lines of curvature (curvature surfaces of dimension one). However, on the two tori, $T^2 \times \{\pm \varepsilon\}$, the principal curvature $\kappa = 0$ has multiplicity two. These two tori are curvature surfaces for this principal curvature, since the principal space corresponding to κ is tangent to each torus at every point. The Legendre submanifold induced from the embedding of M^3 in $\mathbb{R}^4$ has the same properties.

Part (e) of Corollary 4.5 has the following generalization, the proof of which is obtained by invoking the theorem of Ryan [1] mentioned in the proof of Corollary 4.5, with obvious minor modifications.

Corollary 4.7: *Suppose that S is a curvature surface of dimension $m > 1$ in a Legendre submanifold. Then, the corresponding curvature sphere is constant along S.*

A hypersurface $f : M^{n-1} \to S^n$ is called a *Dupin hypersurface* if along each curvature surface, the corresponding principal curvature is constant. We generalize this to the context of Lie sphere geometry by defining a *Dupin submanifold* to be a Legendre submanifold with the property that along each curvature surface, the corresponding curvature sphere is constant. Of course, Legendre submanifolds induced from Dupin hypersurfaces in S^n are Dupin in the sense defined here. But our definition is more general, because the spherical projection need not be an immersion. Corollary 4.7 shows that the only curvature surfaces which must be considered in checking the Dupin condition are those of dimension one. A Dupin submanifold $\lambda : M^{n-1} \to \Lambda^{2n-1}$

is said to be *proper* if the number of distinct curvature spheres is constant on M^{n-1}. The Legendre submanifold induced from the torus T^2 in Example 4.6 above is a proper Dupin submanifold. On the other hand, the Legendre submanifold induced from the tube M^3 over T^2 is Dupin, but not proper. By Theorem 4.3, both the Dupin and proper Dupin properties are invariant under Lie transformations. Because of this, Lie sphere geometry has proven to be a useful setting for the study of Dupin submanifolds.

We now begin the proof of Pinkall's Theorem 4.4. Let $\lambda : M^{n-1} \to \Lambda^{2n-1}$ be a Legendre submanifold with spherical projection f and spherical unit normal field ξ. Given $x \in M^{n-1}$, the differential df is a linear map on $T_x M^{n-1}$ which satisfies

$$df(X) \cdot f(x) = 0 , \qquad df(X) \cdot \xi(x) = 0 ,$$

for all $X \in T_x M^{n-1}$. The second equation holds because λ is a Legendre submanifold. The differential $d\xi$ satisfies

$$d\xi(X) \cdot f(x) = 0 , \qquad d\xi(X) \cdot \xi(x) = 0 ,$$

with the first equation due to the contact condition and the fact that $f \cdot \xi = 0$. Thus, df and $d\xi$ are both linear maps from $T_x M^{n-1}$ to the vector space

$$W_x^{n-1} = (\text{ Span } \{f(x), \xi(x)\} \ )^{\perp} .$$

The first step in the proof of Theorem 4.4 is the following lemma.

Lemma 4.8: *Let* $\lambda : M^{n-1} \to \Lambda^{2n-1}$ *be a Legendre submanifold with spherical projection f and spherical unit normal field ξ. Let x be any point of M^{n-1}. Then, for any two vectors X, Y in $T_x M^{n-1}$, we have*

$$df(X) \cdot d\xi(Y) = df(Y) \cdot d\xi(X) .$$

Proof: Extend X and Y to vector fields in a neighborhood of x such that the Lie bracket $[X, Y] = 0$. First, we differentiate the equation $f \cdot \xi = 0$ and obtain,

$$0 = X(f \cdot \xi) = Xf \cdot \xi + f \cdot X\xi .$$

This, along with the contact condition $(Xf) \cdot \xi = 0$, imply that $f \cdot (X\xi) = 0$. We now differentiate this in the direction of Y to obtain

(4.13) $$Yf \cdot X\xi + f \cdot YX\xi = 0 .$$

Interchanging the roles of X and Y, we also have

(4.14) $$Xf \cdot Y\xi + f \cdot XY\xi = 0 .$$

Since $XY = YX$, we can subtract (4.14) from (4.13) and obtain

$$Xf \cdot Y\xi = Yf \cdot X\xi , \quad \text{i.e.,} \quad df(X) \cdot d\xi(Y) = df(Y) \cdot d\xi(X) . \qquad \square$$

Theorem 4.4 now follows from Lemma 4.9 below with

$$S = df : T_x M^{n-1} \to W_x^{n-1}, \quad T = d\xi : T_x M^{n-1} \to W_x^{n-1} .$$

The linear maps S and T satisfy requirement (a) of Lemma 4.9 because of Lemma 4.8, and they satisfy requirement (b) by the immersion condition in Theorem 2.2. Then by Lemma 4.9, at each $x \in M^{n-1}$, the map

$$df_t = \cos t \, df + \sin t \, d\xi$$

fails to be a bijection for at most $n-1$ values of t in the interval $[0, \pi)$. Thus, Theorem 4.4 is proven. We now prove the key lemma.

Lemma 4.9: *Let V, W be real vector spaces of dimension $n-1$, and suppose that W has a positive definite scalar product (denoted by $\cdot$). Suppose that S and T are linear transformations from V to W which satisfy the following conditions:*
(a) $SX \cdot TY = SY \cdot TX$ *for all $X, Y \in V$.*
(b) *kernel $S \cap$ kernel $T = \{0\}$.*

Then, there are at most $(n - 1 - \dim \ker S)$ values of $a \in \mathbb{R}$ for which the transformation $a\,S + T$ fails to be a bijection.

Proof: Let $V^* = V\,/\,\text{kernel } S$ and $W^* = \text{Image } S$. Then V^* and W^* both have the same dimension, $m = (n - 1 - \dim \ker S)$. For $X \in V$, let X^* denote the image of X under the canonical projection to V^*. For $Y \in W$, let Y^* denote the orthogonal projection of Y onto W^*. Suppose that $Z \in \ker S$. Then for any Y in V, condition (a) implies that

$$TZ \cdot SY = TY \cdot SZ = 0 \ .$$

Thus, TZ is orthogonal to every vector in W^*. Therefore, the mapping,

$$T^* : V^* \to W^* , \qquad T^* X^* = (TX)^* ,$$

is well–defined. Similarly, the mapping $S^* : V^* \to W^*$ given by $S^* X^* = (SX)^*$ is well–defined. Moreover, the map S^* is bijective, since its kernel is 0^*, and V^* and W^* have the same dimension. We can use the bijection S^* to define a positive definite scalar product $(\,,\,)$ on V^* as follows:

$$(X^*, Y^*) = S^* X^* \cdot S^* Y^* \ .$$

Now, define a linear transformation $L : V^* \to V^*$ by $L = S^{*\,-1} T^*$. Then for all X, Y in V, we have

$$
\begin{aligned}
SX \cdot TY &= SX \cdot (TY)^* = S^* X^* \cdot T^* Y^* = S^* X^* \cdot S^* (S^{*\,-1} T^* Y^*) \\
&= (X^*, S^{*\,-1} T^* Y^*) = (X^*, LY^*) \ .
\end{aligned}
$$

Reversing the roles of X and Y, we have

$$SY \cdot TX = (Y^*, LX^*) \ .$$

Thus, L is self–adjoint by (a). Furthermore, for all $X \in V$, we have

(4.15) $\quad TX \cdot TX \geq (TX)^* \cdot (TX)^* = T^*X^* \cdot T^*X^*$
$$= S^*S^{*-1}T^*X^* \cdot S^*S^{*-1}T^*X^* = (LX^*, LX^*) = (X^*, L^2X^*) .$$

Now for $X \in V$ and $a \in \mathbb{R}$, we have $X \in$ kernel $(aS + T)$ if and only if

$$(a\, SX + TX) \cdot (a\, SX + TX) = 0 .$$

Using (4.15), we get

$$(a\, SX + TX) \cdot (a\, SX + TX) = a^2\, SX \cdot SX + 2a\, SX \cdot TX + TX \cdot TX$$
$$\geq a^2\, (X^*, X^*) + 2a\, (X^*, LX^*) + (X^*, L^2X^*)$$
$$= ((a\, I + L)\, X^*, (a\, I + L)\, X^*) \geq 0 ,$$

where I is the identity on V^*. Hence, X is in kernel $(a\, S + T)$ if and only if X^* is in kernel $(a\, I + L)$. Since $a\, I + L$ is a symmetric transformation on a positive definite scalar product space of dimension m, it fails to be a bijection for at most m values of a. For all other values of a, $X \in$ kernel $(a\, S + T)$ implies that $X^* = 0^*$, i.e., $X \in$ kernel S. In that case, the equation,

$$(a\, S + T)\, X = 0 ,$$

implies that $TX = 0$, i.e., $X \in$ kernel T. Then, condition (b) implies that $X = 0$, and thus, $a\, S + T$ is a bijection. $\qquad\qquad \square$

3.5 Lie Curvatures of Legendre Submanifolds

In this section, we introduce certain natural Lie invariants of Legendre submanifolds which have been useful in the classification of Dupin submanifolds.

Let $\lambda : M^{n-1} \to \Lambda^{2n-1}$ be an arbitrary Legendre submanifold. As before, we can write $\lambda = [Y_1, Y_{n+3}]$ with,

(5.1) $\qquad\qquad Y_1 = (1, f, 0) , \quad Y_{n+3} = (0, \xi, 1) ,$

where f and ξ are the spherical projection and spherical field of unit normals, respectively. At each point $x \in M^{n-1}$, the points on the line $\lambda(x)$ can be written in the form,

$$(5.2) \qquad\qquad \mu\, Y_1(x) + Y_{n+3}(x)\ ,$$

i.e., take μ as an inhomogeneous coordinate along the line $\lambda(x)$. Of course, Y_1 corresponds to $\mu = \infty$. The next two propositions give the relationship between the coordinates of the curvature spheres of λ and the principal curvatures of f, in the case where f has constant rank. In the first proposition, we assume that the spherical projection f is an immersion on M^{n-1}. By Theorem 4.4, we know that this can always be achieved locally by passing to a parallel submanifold.

Proposition 5.1: *Let* $\lambda : M^{n-1} \to \Lambda^{2n-1}$ *be a Legendre submanifold whose spherical projection* $f : M^{n-1} \to S^n$ *is an immersion. Let* Y_1 *and* Y_{n+3} *be the point sphere and great sphere maps of* λ *as in* (5.1). *Then, the curvature spheres of* λ *at a point* $x \in M^{n-1}$ *are*

$$[\, K_i\,] = [\kappa_i\, Y_1(x) + Y_{n+3}(x)]\ ,\quad 1 \le i \le g\ ,$$

where $\kappa_1,...,\kappa_g$ *are the distinct principal curvatures at* x *of the oriented hypersurface* f. *The multiplicity of* $[\, K_i\,]$ *equals the multiplicity of* κ_i.

Proof: Let X be a non–zero vector in $T_x M^{n-1}$. Then for any real number μ,

$$d(\,\mu\, Y_1 + Y_{n+3})\,(X) = (0,\, \mu\, df(X) + d\xi(X),\, 0)\ .$$

This is in Span $\{Y_1(x),\, Y_{n+3}(x)\}$ if and only if

$$\mu\, df(X) + d\xi(X) = 0\ ,$$

i.e., μ is a principal curvature of f with corresponding principal vector X. $\square$

A second noteworthy case is when the point sphere map Y_1 is a curvature sphere of constant multiplicity m on M^{n-1}. By Corollary 4.5, the corresponding principal distribution is a foliation, and the curvature sphere $[Y_1]$ is constant along the leaves of this foliation. Thus, the map $[Y_1]$ factors through an immersion $[W_1]$ from the space of leaves V of this foliation into Q^{n+1}. We can write

$$ W_1 = (1, \varphi, 0) \ , $$

where $\varphi : V \to S^n$ is an immersed submanifold of codimension $m + 1$. The manifold M^{n-1} is locally diffeomorphic to an open subset of the unit normal bundle B^{n-1} of the submanifold φ, and λ is essentially the Legendre submanifold induced by φ, as in Section 3.3. The following proposition relates the curvature spheres of λ to the principal curvatures of φ. Recall that the point sphere and great sphere maps for λ are given as in (3.3) by

$$ (5.3) \qquad Y_1(x, \xi) = (1, \varphi(x), 0) \ , \qquad Y_{n+3}(x, \xi) = (0, \xi, 1) \ . $$

Proposition 5.2: *Let* $\lambda : B^{n-1} \to \Lambda^{2n-1}$ *be the Legendre submanifold induced from an immersed submanifold* $\varphi(V)$ *in* S^n *of codimension* $m + 1$. *Let* Y_1 *and* Y_{n+3} *be the point sphere and great sphere maps for* λ *as in* (5.3). *Then, the curvature spheres of* λ *at a point* (x, ξ) *in* B^{n-1} *are*

$$ [K_i] = [\kappa_i Y_1 + Y_{n+3}] \ , \quad 1 \leq i \leq g \ , $$

where $\kappa_1, \ldots, \kappa_{g-1}$ *are the distinct principal curvatures of the shape operator* A^ξ, *and* $\kappa_g = \infty$. *For* $1 \leq i \leq g-1$, *the multiplicity of* $[K_i]$ *equals the multiplicity of* κ_i, *while the multiplicity of* $[K_g]$ *is m.*

Proof: To find the curvature spheres of λ, we use the local trivialization of B^{n-1} given in Section 3.3 and the decomposition of the tangent space to B^{n-1} at (x, ξ) as follows,

$$ T_x V \times \mathrm{Span} \{ \ \partial/\partial t_1 , \ldots , \partial/\partial t_m \ \} = T_x V \times \mathbb{R}^m \ , $$

as in (3.4). First, note that $dY_1(0, Z)$ equals 0 for any $Z \in \mathbb{R}^m$, since Y_1 depends only on x. Hence, Y_1 is a curvature sphere, as expected. Furthermore, since

$$dY_1(X, 0) = (0, d\varphi(X), 0)$$

is never in Span $\{Y_1(x, \xi), Y_{n+3}(x, \xi)\}$ for a non–zero $X \in T_x V$, the multiplicity of the curvature sphere Y_1 is m. If we let $[K_g] = [\kappa_g Y_1 + Y_{n+3}]$ be this curvature sphere, then we must take $\kappa_g = \infty$ to get $[Y_1]$. Using (3.5), we find the other curvature spheres at (x, ξ) by computing

$$d(\mu\, Y_1 + Y_{n+3})\, (X, 0) = (0, d\varphi(\mu\, X - A^\xi X), 0) .$$

From this, it is clear that $[\mu\, Y_1 + Y_{n+3}]$ is a curvature sphere with principal vector $(X, 0)$ if and only if μ is a principal curvature of A^ξ with corresponding principal vector X.   □

Given these two propositions, we define a *principal curvature of a Legendre submanifold* $\lambda : M^{n-1} \to \Lambda^{2n-1}$ at a point $x \in M^{n-1}$ to be a value κ in the set $\mathbb{R} \cup \{\infty\}$ such that $[\kappa\, Y_1(x) + Y_{n+3}(x)]$ is a curvature sphere of λ at x, where Y_1 and Y_{n+3} are as in (5.1).

Remark 5.3: In the case of an immersed submanifold $\varphi(V)$ of codimension $m+1$ in S^n, Reckziegel [2] defines a curvature surface to be a connected submanifold $S \subset V$ for which there is a parallel section $\eta : S \to B^{n-1}$ such that for each $x \in S$, the tangent space $T_x S$ is equal to some smooth eigenspace of $A^{\eta(x)}$. In our formulation, we extend the vector ξ to a parallel normal field ξ_0 in a neighborhood of x, as in Section 3.3. The proof of Proposition 5.2 shows that in our local trivialization of B^{n-1}, any curvature surfaces of the principal curvatures $\kappa_1,...,\kappa_{g-1}$ are of the form $S \times \{0\}$, where $S \subset V$ is a curvature surface in the sense of Reckziegel. In this way, our definition of curvature surface is equivalent to that of Reckziegel. Note that in our formulation, the curvature surfaces for the principal curvature $\kappa_g = \infty$ are of the form $\{x\} \times S^m$. These are not discussed in Reckziegel's approach. Of course, the curvature sphere $[K_g]$ is always constant along these curvature surfaces. Pinkall [5] calls a submanifold $\varphi(V)$ of codimension greater than one "Dupin" if along each curvature surface (in the sense of Reckziegel), the corresponding principal curvature is constant. Thus, the remarks here show that our definition is equivalent to Pinkall's in this situation.

The principal curvatures of a Legendre submanifold are not Lie invariant and depend on the special parametrization (5.1) for λ. However, R. Miyaoka [2] pointed out that the cross–ratios of the principal curvatures are Lie invariant. In order to formulate Miyaoka's theorem, we need to introduce some notation. Suppose that β is a Lie sphere transformation. The Legendre submanifold $\beta\lambda$ has point sphere and great sphere maps given, respectively, by

$$Z_1 = (1, h, 0) , \qquad Z_{n+3} = (0, \zeta, 1) ,$$

where h and ζ are the spherical projection and field of unit normals for $\beta\lambda$. Suppose that

$$[K_i] = [\kappa_i Y_1 + Y_{n+3}] , \quad 1 \le i \le g ,$$

are the distinct curvature spheres of λ at a point $x \in M^{n-1}$. By Theorem 4.3, the points $\beta [K_i]$, $1 \le i \le g$, are the distinct curvature spheres of $\beta\lambda$ at x. We can write

$$\beta [K_i] = [\gamma_i Z_1 + Z_{n+3}] , \quad 1 \le i \le g .$$

These γ_i are the principal curvatures of $\beta\lambda$ at x.

For four distinct numbers a, b, c, d in $\mathbb{R} \cup \{\infty\}$, we adopt the notation,

(5.4) $$[a, b ; c, d] = \frac{(a-b)(d-c)}{(a-c)(d-b)} ,$$

for the cross–ratio of a, b, c, d. Miyaoka's theorem can now be stated as follows.

Theorem 5.4: *Let* $\lambda : M^{n-1} \to \Lambda^{2n-1}$ *be a Legendre submanifold and* β *a Lie sphere transformation. Suppose that* $\kappa_1, ..., \kappa_g$, $g \geq 4$, *are the distinct principal curvatures of* λ *at a point* $x \in M^{n-1}$, *and* $\gamma_1, ..., \gamma_g$ *are the corresponding principal curvatures of* $\beta\lambda$ *at x. Then, for any choice of four numbers h, i, j, k from the set* $\{1, ..., g\}$, *we have*

$$(5.5) \qquad [\kappa_h , \kappa_i ; \kappa_j , \kappa_k] = [\gamma_h , \gamma_i ; \gamma_j , \gamma_k] .$$

Proof: The left side of (5.5) is the cross–ratio, in the sense of projective geometry, of the four points $[K_h]$, $[K_i]$, $[K_j]$, $[K_k]$ on the projective line $\lambda(x)$. The right side of (5.5) is the cross–ratio of the images of these four points under β. The theorem now follows from the fact that the projective transformation β preserves the cross–ratio of four points on a line. □

The cross–ratios of the principal curvatures of λ are called the *Lie curvatures* of λ. A set of related invariants for the Moebius group is obtained as follows. First, recall that a Moebius transformation is a Lie transformation which takes point spheres to point spheres. Hence, the transformation β in Theorem 5.4 is a Moebius transformation if and only if $\beta [Y_1] = [Z_1]$. This leads to the following corollary of Theorem 5.4.

Corollary 5.5: *Let* $\lambda : M^{n-1} \to \Lambda^{2n-1}$ *be a Legendre submanifold, and let* β *be a Moebius transformation. Then, for any three distinct principal curvatures* κ_h , κ_i, κ_j *of* λ *at* $x \in M^{n-1}$, *none of which equals* ∞ , *we have*

$$(5.6) \quad \Phi (\kappa_h , \kappa_i , \kappa_j) = (\kappa_h - \kappa_i) / (\kappa_h - \kappa_j) = (\gamma_h - \gamma_i) / (\gamma_h - \gamma_j) ,$$

where γ_h , γ_i *and* γ_j *are the corresponding principal curvatures of* $\beta\lambda$ *at the point x.*

Proof: First, note that we are using (5.6) to define the quantity Φ. Now, since β is a Moebius transformation, the point $[Y_1]$, corresponding to $\mu = \infty$, is taken by β to the point $[Z_1]$ with coordinate $\gamma = \infty$. Since β preserves cross–ratios, we have

$$[\; \kappa_h \, , \kappa_i \, ; \kappa_j \, , \infty \;] = [\; \gamma_h \, , \gamma_i \, ; \gamma_j \, , \infty \;] \; .$$

The corollary now follows, since the cross–ratio on the left equals the left side of (5.6), and the cross–ratio on the right equals the right side of (5.6). □

A ratio Φ of the form (5.6) is called a *Moebius curvature* of λ. Lie and Moebius curvatures have already been useful in characterizing which Legendre submanifolds are Lie equivalent to Legendre submanifolds induced from isoparametric hypersurfaces in spheres. Recall that an immersed hypersurface $f : M^{n-1} \to S^n$ is called *isoparametric* if f has constant principal curvatures. The study of isoparametric hypersurfaces in spheres was begun by E. Cartan [2]–[5], and further developed by Nomizu [3]–[4], Takagi and Takahashi [1], Ozeki and Takeuchi [1], and most extensively by Münzner [1], who showed that the number g of distinct principal curvatures of an isoparametric hypersurface must be 1, 2, 3, 4 or 6. (See Chapter 3 of Cecil–Ryan [7] for more detail.) Cartan classified isoparametric hypersurfaces with $g = 1$, 2 or 3 principal curvatures, but the cases $g = 4$ and 6 have not yet been completely classified. Ferus, Karcher and Münzner [1] constructed a large class of examples with $g = 4$ using representations of Clifford algebras. Their construction included all previously known examples, except two. Later, Pinkall and Thorbergsson [1] gave an alternate geometric construction of the Clifford examples (see Section 3.7). Wang [1]–[2] provided more information on isoparametric hypersurfaces and the topology of the Clifford examples.

Abresch [1] discovered some restrictions on the multiplicities of the principal curvatures when $g = 4$ or 6. In particular, for the case $g = 6$, all the multiplicities must be equal, and M must have dimension 6 or 12. Dorfmeister and Neher [1] proved that if dim $M = 6$, then M must be homogeneous. It remains an open question whether this is also true if dim $M = 12$. Finally, Solomon [1] has begun a study of some aspects of the intrinsic Riemannian geometry of isoparametric hypersurfaces.

There is also a theory of isoparametric submanifolds of codimension greater than one, due principally to Carter and West [2]–[4], Terng [1], and Hsiang, Palais, and Terng [1]. (See also Harle [1] and Strübing [1].) A connected, complete submanifold V in a real space–form is said to be *isoparametric* if it has flat normal bundle and if for any parallel section of the unit normal bundle $\eta : V \rightarrow B(V)$, the principal curvatures of A^η are constant. This clearly implies that V is Dupin, since the Dupin condition only requires that the principal curvatures are constant along their corresponding curvature surfaces (See Remark 5.3.). Recently, a beautiful result of Thorbergsson [2] has shown that the isoparametric submanifolds of codimension greater than one in a sphere are all homogeneous. See Heintze, Olmos and Thorbergsson [1] for further results in this area.

Terng [4] introduced a theory of isoparametric submanifolds in infinite dimensional spaces. Pinkall and Thorbergsson [2] then provided more examples of such submanifolds.

Münzner's work shows that any connected piece of isoparametric hypersurface in S^n can be extended to a compact isoparametric hypersurface in a unique way. The following is a local Lie geometric characterization of those Legendre submanifolds which are Lie equivalent to Legendre submanifolds induced by isoparametric hypersurfaces in spheres (see Cecil [3]). Recall that a line in $\mathbb{P}^{n+2}$ is called *timelike* if it contains only timelike points. This means that an orthonormal basis for the 2–plane in $\mathbb{R}_2^{n+3}$, determined by the timelike line, consists of two timelike vectors. An example is the line $[e_1, e_{n+3}]$.

Theorem 5.6: *Let* $\lambda : M^{n-1} \rightarrow \Lambda^{2n-1}$ *be a Legendre submanifold with* g *distinct curvature spheres* $[K_1],...,[K_g]$ *at each point. Then* λ *is Lie equivalent to the Legendre submanifold induced by an isoparametric hypersurface in* S^n *if and only if there exists* g *points* $[P_1],...,[P_g]$ *on a timelike line in* $\mathbb{P}^{n+2}$ *such that*

$$< K_i , P_i > = 0 , \quad 1 \le i \le g .$$

Proof: If λ is the Legendre submanifold induced by an isoparametric hypersurface, then all the spheres in a family $[K_i]$ of curvature spheres have the same radius ρ_i, where $0 < \rho_i < \pi$. By formula (4.4) of Chapter 1, this is equivalent to the condition $< K_i , P_i > = 0$, where

(5.7) $\qquad P_i = \sin \rho_i \; e_1 - \cos \rho_i \; e_{n+3} , \quad 1 \le i \le g ,$

are g points on the timelike line $[e_1, e_{n+3}]$. Since a Lie transformation preserves curvature spheres, timelike lines and the polarity relationship, the same is true for any Lie image of λ.

Conversely, suppose that there exist points $[P_1],...,[P_g]$ on a timelike line ℓ such that $< K_i, P_i > = 0$. Let β be a Lie transformation which takes ℓ to the line $[e_1, e_{n+3}]$. Then the curvature spheres $\beta [K_i]$ of $\beta\lambda$ are respectively orthogonal to the points $[Q_i] = \beta [P_i]$ on the line $[e_1, e_{n+3}]$. This means that the spheres corresponding to $\beta [K_i]$ have constant radius on M^{n-1}. By applying a parallel transformation, if necessary, we can arrange that none of these curvature spheres has radius zero. Then, $\beta\lambda$ is the Legendre submanifold induced from an isoparametric hypersurface in S^n.   □

Remark 5.7: In the case where λ is Lie equivalent to an isoparametric hypersurface, one can say more about the position of the points $[P_i]$ along the line ℓ. Münzner [1] showed that the radii ρ_i of the curvature spheres of an isoparametric hypersurface must be of the form

$$(5.8) \qquad \rho_i = \rho_1 + (i-1)\, \pi\, /\, g\, , \qquad 1 \le i \le g\, ,$$

for some $\rho_1 \in (0, \pi/g)$. Hence, after Lie transformation, the $[P_i]$ must have the form (5.7), for ρ_i as in (5.8).

Since the principal curvatures are constant on an isoparametric hypersurface, the Lie curvatures are also constant. By Münzner's work, the distinct principal curvatures κ_i, $1 \le i \le g$, of an isoparametric hypersurface must have the form

$$(5.9) \qquad \kappa_i = \cot \rho_i\, ,$$

for ρ_i as in (5.8). Thus, the Lie curvatures of an isoparametric hypersurface can be determined. We can order the principal curvatures,

$$\kappa_1 < \ldots < \kappa_g\, .$$

In the case $g = 4$, this leads to a unique Lie curvature Ψ defined by

(5.10) $\Psi = [\ \kappa_1, \kappa_2\ ;\ \kappa_3, \kappa_4\ ] = (\kappa_1-\kappa_2)(\kappa_4-\kappa_3)/(\kappa_1-\kappa_3)(\kappa_4-\kappa_2)$.

The ordering of the principal curvatures implies that Ψ must satisfy $0 < \Psi < 1$. Using (5.9) and (5.10), one can compute that $\Psi = 1/2$ on any isoparametric hypersurface, i.e., the four curvature spheres form a harmonic set, in the sense of projective geometry (see, for example, Samuel [1, p.59]). There is, however, a simpler way to compute Ψ. One applies Proposition 5.2 to the Legendre submanifold induced by one of the focal submanifolds of the isoparametric hypersurface. By the work of Münzner, each isoparametric hypersurface M^{n-1} embedded in S^n has two distinct focal submanifolds, each of codimension greater than one. M^{n-1} is a tube of constant radius over each of these submanifolds. Therefore, the Legendre submanifold induced by M^{n-1} is obtained from the Legendre submanifold induced by a focal submanifold by a parallel transformation. Thus, these two Legendre submanifolds have the same Lie curvature. Let $\varphi : V \to S^n$ be one of these focal submanifolds. By the same calculation which yields (5.8), Münzner showed that if ξ is any unit normal to $\varphi(V)$ at any point, then the shape operator A^ξ has three distinct principal curvatures, $\kappa_1 = -1$, $\kappa_2 = 0$, $\kappa_3 = 1$. By Proposition 5.2, the Legendre submanifold induced by φ has a fourth principal curvature $\kappa_4 = \infty$. Thus, the Lie curvature of this Legendre submanifold is

$$\Psi = (-1 - 0)(\infty - 1) / (-1 - 1)(\infty - 0) = 1 / 2\ .$$

We can determine the Lie curvatures of an isoparametric hypersurface M^{n-1} in S^n with $g = 6$ principal curvatures in the same way. Let $\varphi(V)$ be one of the focal submanifolds of M^{n-1}. By Münzner's formula (5.8), the Legendre submanifold induced from φ has six constant principal curvatures,

$$\kappa_1 = -\sqrt{3}\ ,\quad \kappa_2 = -1/\sqrt{3}\ ,\quad \kappa_3 = 0\ ,\quad \kappa_4 = 1/\sqrt{3}\ ,\quad \kappa_5 = \sqrt{3}\ ,\quad \kappa_6 = \infty\ .$$

The corresponding six curvature spheres are situated symmetrically, as in Figure 5.1. There are only three geometrically distinct configurations which can be obtained by choosing four of the six curvature spheres. These give the cross–ratios:

$$[\kappa_3, \kappa_4; \kappa_5, \kappa_6] = 1/3 \ , \quad [\kappa_2, \kappa_3; \kappa_5, \kappa_6] = 1/4 \ , \quad [\kappa_2, \kappa_3; \kappa_4, \kappa_6] = 1/2 \ .$$

Of course, if a certain cross–ratio has the value Ψ, then one can obtain the values,

$$\{ \ \Psi \ , \ 1/\Psi \ , \ 1-\Psi \ , \ 1/(1-\Psi) \ , \ (\Psi-1)/\Psi \ , \ \Psi/(\Psi-1) \ \} \ ,$$

by permuting the order of the spheres (see, for example, Samuel [1, p.58]).

Returning to the case $g = 4$, one can ask what is the strength of the assumption $\Psi = 1/2$. Since Ψ is only one function of the principal curvatures, one would not expect this assumption to classify Legendre submanifolds up to Lie equivalence. However, if one makes additional assumptions, e.g., the Dupin condition, then results can be obtained. Specifically, Miyaoka [2] proved that the assumption that Ψ is constant on a compact embedded proper Dupin hypersurface M^{n-1} in S^n, together with an additional assumption regarding intersections of leaves of the principal foliations, implies that M^{n-1} is Lie equivalent to an isoparametric hypersurface. Whether this additional assumption can be dropped is an open question.

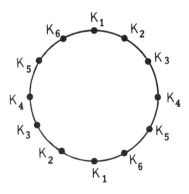

Figure 5.1 – Curvature spheres on a projective line, $g = 6$

The following example (Cecil [3]) shows that some global hypotheses are necessary to reach Miyaoka's conclusion. This example is a non–compact proper Dupin submanifold with $g = 4$ distinct curvature spheres and $\Psi = 1/2$, which is not Lie equivalent to an open subset of an isoparametric hypersurface. This example cannot be made compact without destroying the property that $g = 4$ at each point. Let $\varphi : V \to S^{n-m}$ be an embedded Dupin hypersurface in S^{n-m} with field of unit normals ξ, such that φ has three distinct principal curvatures $\mu_1 < \mu_2 < \mu_3$ at each point of V. Embed S^{n-m} as a totally geodesic submanifold in S^n, and let B^{n-1} be the unit normal bundle of the submanifold $\varphi(V)$ in S^n. Let $\lambda : B^{n-1} \to \Lambda^{2n-1}$ be the Legendre submanifold induced from the submanifold $\varphi(V)$ in S^n. Any unit normal η to $\varphi(V)$ at a point $x \in V$ can be written as

$$\eta = \cos \theta \, \xi(x) + \sin \theta \, \zeta \, ,$$

where ζ is a unit normal to S^{n-m} in S^n. Since the shape operator $A^\zeta = 0$, we have

$$A^\eta = \cos \theta \, A^\xi \, .$$

Thus, the principal curvatures of A^η are

(5.11) $\kappa_i = \cos \theta \, \mu_i, \quad 1 \leq i \leq 3 \, .$

If $\eta \cdot \xi = \cos \theta \neq 0$, then A^η has three distinct principal curvatures. However, if $\eta \cdot \xi = 0$, then $A^\eta = 0$. Let U be the open subset of B^{n-1} on which $\cos \theta > 0$, and let α denote the restriction of λ to U. By Proposition 5.2, α has four distinct curvature spheres at each point of U. Since $\varphi(V)$ is Dupin in S^{n-m}, it is easy to show that α is Dupin (see the tube construction in Section 4.2 for the details). Furthermore, since $\kappa_4 = \infty$, the Lie curvature of α at a point (x, η) of U equals the Moebius curvature $\Phi(\kappa_1, \kappa_2, \kappa_3)$. Using (5.11), we compute

(5.12) $\Psi(x, \eta) = \Phi(\kappa_1, \kappa_2, \kappa_3) = (\mu_1 - \mu_2) / (\mu_1 - \mu_3) = \Phi(\mu_1, \mu_2, \mu_3) \, .$

Now, suppose that $\varphi(V)$ is a minimal isoparametric hypersurface in S^{n-m} with three distinct principal curvatures. By Münzner's formula (5.8), these principal curvatures must have the values $\mu_1 = -\sqrt{3}$, $\mu_2 = 0$, $\mu_3 = \sqrt{3}$. On the open subset U of B^{n-1} described above, the Lie curvature of α has the constant value 1/2 by (5.12). To construct an immersed proper Dupin hypersurface with 4 principal curvatures and $\Psi = 1/2$ in S^n, we simply take the open subset $\varphi_t(U)$ of the tube of radius t around $\varphi(V)$ in S^n.

It is not hard to see that this example is not Lie equivalent to an open subset of an isoparametric hypersurface with four distinct principal curvatures in S^n. Note that the point sphere map $[Y_1]$ of α is a curvature sphere of multiplicity m, which lies in the linear subspace of codimension $m+1$ in $\mathbb{P}^{n+2}$ orthogonal to the space spanned by e_{n+3} and by those vectors ζ normal to S^{n-m} in S^n. This geometric fact implies that for such ζ, there are only two distinct curvature spheres on each of the lines $\lambda(x, \zeta)$, since $A^\zeta = 0$ (see Proposition 5.2). On the other hand, if $\gamma : M^{n-1} \to \Lambda^{2n-1}$ is the Legendre submanifold induced by an isoparametric hypersurface with four principal curvatures in S^n, then there are four distinct curvature spheres on each line $\gamma(x)$, for $x \in M^{n-1}$. Thus, no curvature sphere of γ lies in a linear subspace of codimension greater than one in $\mathbb{P}^{n+2}$, and γ is not Lie equivalent to α. This change in the number of distinct curvature spheres at points of the form (x, ζ) is precisely why α cannot be extended to a compact proper Dupin submanifold with $g = 4$.

With regard to Theorem 5.6, α comes as close as possible to satisfying the requirements for being Lie equivalent to an isoparametric hypersurface without actually fulfilling them. The principal curvatures $\kappa_2 = 0$ and $\kappa_4 = \infty$ are constant on U. If a third principal curvature were also constant, then the constancy of Ψ would imply that all four principal curvatures were constant, and α would be isoparametric.

Using this same method, it is easy to construct non–compact proper Dupin hypersurfaces in S^n with $g = 4$ and $\Psi = c$, for any constant $0 < c < 1$. If $\varphi(V)$ is an isoparametric hypersurface in S^{n-m} with 3 principal curvatures, then Münzner's formula (5.8) implies that these must have the values,

(5.13) $\mu_1 = \cot(\theta + (2\pi/3))$, $\mu_2 = \cot(\theta + (\pi/3))$, $\mu_3 = \cot\theta$, $0 < \theta < \pi/3$.

Furthermore, any value of θ in $(0, \pi/3)$ can be realized by some hypersurface in a parallel family of isoparametric hypersurfaces. A direct calculation using (5.12) and (5.13) shows that the Lie curvature Ψ of α satisfies,

$$\Psi = \Phi\left(\kappa_1, \kappa_2, \kappa_3\right) = \frac{\mu_1 - \mu_2}{\mu_1 - \mu_3} = \frac{1}{2} + \frac{\sqrt{3}}{2} \tan\left(\theta - (\pi/6)\right),$$

on U. This can assume any value c in $(0, 1)$ by an appropriate choice of θ in $(0, \pi/3)$. An open subset of a tube over $\varphi(V)$ in S^n is a proper Dupin hypersurface with $g = 4$ and $\Psi = \Phi = c$. Since Φ has different values on different hypersurfaces in the parallel family, these hypersurfaces are not Moebius equivalent by Corollary 5.5. Of course, this can also be deduced from the fact that parallel transformation is not a Moebius transformation.

3.6 Taut Submanifolds

In this section, we prove that tautness is invariant under Lie sphere transformations. First, we briefly review the definition and basic facts concerning taut immersions into real space forms. The reader is referred to Chapter 2 of Cecil–Ryan [7] for more detail and additional references.

Let $\varphi : V \to \mathbb{R}^n$ be an immersion of a compact connected manifold V into $\mathbb{R}^n$. For p in $\mathbb{R}^n$, the Euclidean distance function $L_p : V \to \mathbb{R}$ is defined by the formula $L_p(x) = |p - \varphi(x)|^2$. If p is not a focal point of the submanifold φ, then L_p is *nondegenerate*, i.e., all of its critical points are nondegenerate (see Milnor [1, p.32–38]). By the Morse inequalities, the number $\mu(L_p)$ of critical points of a nondegenerate distance function satisfies $\mu(L_p) \geq \beta(V;\mathbb{F})$, the sum of the $\mathbb{F}$–Betti numbers of V for any field $\mathbb{F}$. The immersion φ is said to be *taut* if there exists a field $\mathbb{F}$ such that every nondegenerate Euclidean distance function has $\beta(V;\mathbb{F})$ critical points on V. If it is necessary to distinguish the field $\mathbb{F}$, we will say that φ is $\mathbb{F}$–*taut*. The field $\mathbb{F} = \mathbb{Z}_2$ has been satisfactory for most considerations thus far.

Tautness was first studied by Banchoff [1], who determined all taut surfaces in Euclidean space. Carter and West [1] proved many basic results about taut immersions. Among these is the fact that a taut immersion must be an embedding, since if $p \in \mathbb{R}^n$ were a double point, then the function L_p would have two absolute minima.

Tautness is invariant under Moebius transformations of $\mathbb{R}^n \cup \{\infty\}$. Further, an embedding $\varphi:V \to \mathbb{R}^n$ is taut if and only if the embedding $\sigma\varphi:V \to S^n$ has the property that every nondegenerate spherical distance function has $\beta(V;\mathbb{F})$ critical points on V, where $\sigma : \mathbb{R}^n \to S^n - \{P\}$ is stereographic projection (see Section 1.2). Since a spherical distance function is essentially a Euclidean height function $\ell_p(q) = p \cdot q$, for $p, q \in S^n$, the embedding φ is taut if and only if the spherical embedding $\sigma\varphi$ is *tight*, i.e., every nondegenerate height function ℓ_p has $\beta(V;\mathbb{F})$ critical points on V. It is also easy to show that a tight spherical embedding $\psi : V \to S^n \subset \mathbb{R}^{n+1}$ is taut when considered as an embedding of V into $\mathbb{R}^{n+1}$ (see Cecil–Ryan [7, p.114]).

Tight and taut submanifolds have also been studied in hyperbolic space H^n. Cecil and Ryan [3]–[4] defined three classes of distance functions in H^n whose level sets are respectively: spheres centered at a point p in H^n, equidistant hypersurfaces from a hyperplane π in H^n, and horospheres equidistant from a fixed horosphere h. Now, suppose that $V \subset H^n$ is a compact connected submanifold embedded in H^n and that $\tau : H^n \to D^n$ is stereographic projection of H^n onto a disk D^n in $\mathbb{R}^n$, as in (4.8) of Chapter 1. Then $\tau(V) \subset \mathbb{R}^n$ is taut if and only if every nondegenerate hyperbolic distance function of each of the three types has $\beta(V;\mathbb{F})$ critical points on V (see Cecil–Ryan [4, p.563]).

As we saw in Section 3.3, any immersed hypersurface $f : M^{n-1} \to S^n$ naturally induces a Legendre submanifold $\lambda : M^{n-1} \to \Lambda^{2n-1}$. Further, any submanifold V of codimension greater than one in a real space form induces a Legendre submanifold defined on the unit normal bundle $B(V)$ of V. This Legendre submanifold is Lie equivalent to the Legendre submanifold induced by a tube of constant radius ε over V. In the theory of taut embeddings with coefficient field $\mathbb{F} = \mathbb{Z}_2$, it also suffices to a large extent to work with hypersurfaces for the following reason. Let $V \subset S^n$ be a compact connected submanifold of dimension $k < n-1$, and let T be a tube of sufficiently small radius $\varepsilon > 0$ around V so that T is an embedded hypersurface in S^n. Then, Pinkall [5] showed that V is taut with $\mathbb{F} = \mathbb{Z}_2$ if and only if T is taut. One computes directly that if p is a point in S^n but not in V, then ℓ_p is nondegenerate on T if and only if it is nondegenerate on V. Furthermore, the number of critical points of ℓ_p on T is always twice the number of critical points of ℓ_p on V. Pinkall used the Gysin sequence for $B(V)$ to show that $\beta(T;\mathbb{Z}_2) = 2\beta(V;\mathbb{Z}_2)$, thereby completing the proof of his result.

Thus, in formulating our first result on tautness for Legendre submanifolds, we will assume that the Moebius projection is an immersion. We now briefly recall the notion of a Moebius projection from Section 3.3. Let $\lambda : M^{n-1} \to \Lambda^{2n-1}$ be a Legendre submanifold. For each $x \in M^{n-1}$, the parabolic pencil determined by the line $\lambda(x)$ contains exactly one point sphere $[Y_1(x)]$, corresponding to the point of intersection of $\lambda(x)$ with the hyperplane $e_{n+3}^{\perp}$. The map $[Y_1]$ from M^{n-1} into Q^{n+1} is called the Moebius projection of λ. This map is an immersion at x if and only if $[Y_1(x)]$ is not a curvature sphere of λ at x.

If we choose an orthonormal basis $\{e_1,...,e_{n+2}\}$ for $e_{n+3}^{\perp}$, then $[Y_1]$ determines a spherical projection f of M^{n-1} into the unit sphere S^n in the Euclidean space $\mathbb{R}^{n+1}$ spanned by $\{e_2,...,e_{n+2}\}$, by the formula $Y_1 = (1, f, 0)$. The map $[Y_1]$ is an immersion if and only if f is an immersion. This fact is independent of the choice of basis of $e_{n+3}^{\perp}$, i.e., of the choice of spherical projection. Moreover, $[Y_1]$ is an immersion if and only if any Euclidean or hyperbolic projection of λ is an immersion. If the Moebius projection $[Y_1]$ is an immersion, we say that $[Y_1]$ is *taut* if some spherical projection f of λ is taut. This is also independent of the choice of spherical projection, since all spherical projections are Moebius equivalent, and tautness is invariant under Moebius transformations. Tautness of $[Y_1]$ is also equivalent to the tautness of some Euclidean or hyperbolic projection of λ by the remarks above.

Theorem 6.1: *Let* $\lambda : M^{n-1} \to \Lambda^{2n-1}$ *be a compact, connected Legendre submanifold whose Moebius projection is a taut immersion. Suppose* β *is a Lie transformation such that the Moebius projection of* $\beta\lambda$ *is an immersion. Then the Moebius projection of* $\beta\lambda$ *is taut.*

Proof: The theorem follows almost immediately from Theorem 5.3 of Chapter 2 which states that β can be written in the form $\Phi P_t \Psi$, where Φ and Ψ are Moebius transformations and P_t is some Euclidean, spherical or hyperbolic parallel transformation. By hypothesis, the Moebius projection of λ is an immersion. Since this property is preserved by Moebius transformations, the assumption that the Moebius projection of $\beta\lambda$ is an immersion is equivalent to assuming that the Moebius projection of $P_t\,\mu$ is an immersion, where $\mu = \Psi\lambda$. Finally, since tautness is preserved by Moebius transformations, we can complete the proof by showing that the Moebius projection of $P_t\,\mu$ is taut. We may arrange by an appropriate choice of basis vectors $e_1,...,e_{n+2}$ that the Moebius projection of μ does not meet the improper point $[e_1 - e_2]$. Then with respect to this basis, the Moebius projection Z_1 of μ takes the form

(6.1) $Z_1 = (\, 1 + |F|^2\,,\, 1 - |F|^2\,,\, 2F,\, 0\,)\,/\,2\,,$

where $F : M^{n-1} \to \mathbb{R}^n = \mathrm{Span}\ \{e_3,...,e_{n+2}\}$ is the Euclidean projection of μ . We first consider the case of a Euclidean P_t . A calculation similar to (4.12) shows that the Euclidean projection of $P_t\,\mu$ is the parallel hypersurface F_{-t} at oriented distance $-t$ from F. Since an immersed parallel hypersurface to a taut hypersurface is taut (see Cecil–Ryan [7, p.185]), F_{-t} is taut, and so the Moebius projection of $P_t\,\mu$ is taut. The same argument applied to the spherical projection of μ shows that the Moebius projection of $P_t\,\mu$ is taut for a spherical parallel transformation. Finally, suppose that P_t is a hyperbolic parallel transformation. Since M^{n-1} is compact, there is some $\rho > 0$ such that $|F| < \rho$ on M^{n-1}. We can place a hyperbolic metric g on the disk D^n of radius ρ centered at the origin so that P_t acts as a hyperbolic parallel transformation on $H^n = (D^n, g)$. The Euclidean tautness of F is equivalent to the condition that every nondegenerate hyperbolic distance function of each of the three types has $\beta(M^{n-1};\mathbb{F})$ critical points on M^{n-1}. The hyperbolic projection of $P_t\,\mu$ is the parallel hypersurface h_{-t} at oriented distance $-t$ from F with respect to the hyperbolic metric g on D^n. By Theorem C of Cecil–Ryan [3], h_{-t} also has the property that every nondegenerate function of each of the three types has $\beta(M^{n-1};\mathbb{F})$ critical points on M^{n-1}. Hence, in the Euclidean metric on D^n, the hypersurface h_{-t} is taut, and so the Moebius projection of $P_t\,\mu$ is taut, as desired.   □

We can now formulate two corollaries of this theorem in terms of submanifolds of S^n. Two hypersurfaces $f : M^{n-1} \to S^n$ and $g : M^{n-1} \to S^n$ are said to be Lie equivalent if their induced Legendre submanifolds λ and μ are Lie equivalent, i.e., there exists a Lie sphere transformation β such that $\mu = \beta\lambda$.

Corollary 6.2: *Let f and g be two Lie equivalent immersions of a compact, connected manifold M^{n-1} into S^n. Then f is taut if and only if g is taut.*

Proof: Since f and g are both immersions, the Moebius projections of their induced Legendre submanifolds λ and $\mu = \beta\lambda$ are immersions. The immersion f is taut if and only if the Moebius projection of λ is taut. By Theorem 6.1, this implies that the Moebius projection of μ is taut, which in turn implies that g is taut. Since the roles of f and g can be reversed, the result is proven. $\square$

Finally, we use Pinkall's theorem concerning $\mathbb{Z}_2$–tautness to extend these results to submanifolds of higher codimension. As we saw in Section 3.3, any submanifold V embedded in S^n induces a Legendre submanifold λ defined on the unit normal bundle $B(V)$ in S^n. As with hypersurfaces, two submanifolds V and W of arbitrary codimension in S^n are said to be Lie equivalent if their induced Legendre submanifolds are Lie equivalent.

Corollary 6.3: *Let V and W be two Lie equivalent compact, connected submanifolds embedded in S^n. Then V is $\mathbb{Z}_2$-taut if and only if W is $\mathbb{Z}_2$-taut.*

Proof: Let $\hat{V}$ and $\hat{W}$ be tubes of sufficiently small radius over V and W, respectively, that they are embedded hypersurfaces in S^n. The Lie equivalence of V and W implies that $\hat{V}$ and $\hat{W}$ are Lie equivalent, since they are equivalent by parallel transformation to V and W, respectively. Now, assume that V is $\mathbb{Z}_2$–taut. Then $\hat{V}$ is $\mathbb{Z}_2$–taut by Pinkall's theorem. Next, Corollary 6.2 implies that $\hat{W}$ is $\mathbb{Z}_2$–taut. Finally, W is $\mathbb{Z}_2$–taut by Pinkall's theorem. As before, the roles of V and W can be reversed. $\square$

As we noted in the Introduction, Pinkall [5] showed that taut implies Dupin (see also Cecil–Ryan [7, p.195]). Thorbergsson [1] showed that a compact proper Dupin hypersurface embedded in $\mathbb{R}^n$ is taut. An important open question is whether Dupin implies taut without the assumption that the number of distinct principal curvatures is constant.

It is well known that the compact focal submanifolds of a taut hypersurface in $\mathbb{R}^n$ need not be taut. For example, one focal submanifold of a non–round cyclide of Dupin in $\mathbb{R}^3$ is an ellipse. This is tight but not taut. More generally, Buyske [1] has used Lie sphere–geometric techniques to show that if a hypersurface M^{n-1} in $\mathbb{R}^n$ is Lie equivalent to an isoparametric hypersurface in S^n, then each compact focal submanifold of M^{n-1} is tight in $\mathbb{R}^n$.

3.7 Compact Proper Dupin Submanifolds

In this section, we discuss compact proper Dupin submanifolds. The first examples are those which are Lie equivalent to an isoparametric hypersurface in the sphere S^n. Münzner [1] showed that the number g of distinct principal curvatures of an isoparametric hypersurface in S^n must be 1, 2, 3, 4 or 6. Thorbergsson [1] then showed that the same restriction holds for a compact proper Dupin hypersurface M^{n-1} embedded in S^n. He first proved that M^{n-1} must be taut in S^n. Using the tautness, he then showed that M^{n-1} divides S^n into two ball bundles over the first focal submanifolds on either side of M^{n-1} in S^n. This topological data is all that is required for Münzner's restriction on g.

Compact proper Dupin hypersurfaces in S^n have been classified in the cases $g = 1$, 2 and 3. In each case, M^{n-1} must be Lie equivalent to an isoparametric hypersurface. The case $g = 1$ is simply the case of umbilic hypersurfaces, i.e., hyperspheres in S^n. In the case $g = 2$, Cecil and Ryan [2] showed that M^{n-1} must be a cyclide of Dupin (see Section 4.4), and thus, it is Moebius equivalent to a standard product of spheres $S^k \times S^{n-1-k}$ in S^n. In the case $g = 3$, Miyaoka [1] proved that M^{n-1} must be Lie equivalent to an isoparametric hypersurface. Earlier, Cartan [3] had shown that an isoparametric hypersurface with $g = 3$ is a tube over a standard Veronese embedding of a projective plane $\mathbb{F}P^2$, for $\mathbb{F} = \mathbb{R}, \mathbb{C}, \mathbb{H}$ (quaternions) or O (Cayley numbers), in S^4, S^7, S^{13} and S^{25}, respectively. These results led to the widely held conjecture that every compact proper Dupin hypersurface embedded in S^n is Lie equivalent to an isoparametric hypersurface (see Cecil–Ryan [7, p.184]). All attempts to verify the conjecture in the cases $g = 4$ and 6 were unsuccessful. Finally, in 1988 Pinkall and Thorbergsson [1] and Miyaoka and

Ozawa [1] gave two different methods for producing counterexamples to the conjecture with $g = 4$. The method of Miyaoka and Ozawa also yields counterexamples in the case $g = 6$. In this section, we will present these counterexamples. In both cases, the proof that the examples are not Lie equivalent to an isoparametric hypersurface is accomplished by showing that the Lie curvature Ψ does not have the constant value 1/2 (see Section 3.5).

The construction of Pinkall and Thorbergsson begins with an isoparametric hypersurface with four principal curvatures, or rather with one of its focal submanifolds. In the first example of this type, due to Cartan [5], the focal submanifold is the Stiefel manifold of orthonormal 2–frames in $\mathbb{R}^3$. This was generalized by Nomizu [3]–[4] to the case where the focal submanifold is the Stiefel manifold $V_{n+1,2}(\mathbb{R})$ of orthonormal 2–frames in $\mathbb{R}^{n+1}$, $n \geq 2$. Later Ferus, Karcher, and Münzner [1], using representations of Clifford algebras, gave a construction which yields all known examples of isoparametric hypersurfaces with $g = 4$, except two. Then, Pinkall and Thorbergsson pointed out that for the examples of Ferus, Karcher and Münzner, one focal submanifold is a Clifford–Stiefel manifold, a generalization of $V_{n+1,2}(\mathbb{R})$ to the appropriate Clifford algebra. We will treat the real case here. The details of the generalization to Clifford algebras are presented clearly in the paper of Pinkall and Thorbergsson.

Let S^{2n+1} be the unit sphere in $\mathbb{R}^{2n+2} = \mathbb{R}^{n+1} \times \mathbb{R}^{n+1}$. We consider the Stiefel manifold,

$$V = \{(x, y) \in \mathbb{R}^{n+1} \times \mathbb{R}^{n+1} \mid |x| = |y| = 1 / \sqrt{2}, \ x \cdot y = 0\} \ .$$

The submanifold V of codimension two in S^{2n+1} induces a Legendre submanifold defined on the unit normal bundle $B(V)$ of V in S^{2n+1}. We will show that if ζ is any unit normal to V at any point, then the shape operator A^ζ has three distinct principal curvatures,

(7.1) $\kappa_1 = -1 , \quad \kappa_2 = 0 , \quad \kappa_3 = 1 ,$

with respective multiplicities $n-1$, 1, $n-1$. As in Proposition 5.2, the Legendre submanifold has a fourth principal curvature $\kappa_4 = \infty$ of multiplicity 1 at each point of $B(V)$. Since $\kappa_4 = \infty$, the Lie curvature Ψ at any point of $B(V)$ equals the Moebius curvature Φ, i.e.,

$$\Psi = \Phi = (\kappa_1 - \kappa_2) / (\kappa_1 - \kappa_3) = (-1 - 0) / (-1 - 1) = 1 / 2 \; .$$

Since all four principal curvatures are constant on $B(V)$, a tube over V in S^{2n+1} is an isoparametric hypersurface with four principal curvatures (see, for example, Cecil–Ryan [7, p.131]).

We now show that the principal curvatures of V are as in (7.1). Note that

$$V \subset f^{-1}(0) \cap g^{-1}(0) \; ,$$

where f and g are the real valued functions defined on S^{2n+1} by

(7.2) $f(u, v) = (v \cdot v - u \cdot u) / 2 \; , \quad g(u, v) = - u \cdot v \; .$

Thus, the gradients,

$$\xi = (-u, v) \; , \quad \eta = (-v, -u) \; ,$$

of f and g are two fields of unit normals to V in S^{2n+1}. Moreover, they are orthogonal to each other.

Let (u, v) be an arbitrary point in V. Let $e_1,...,e_{n-1}$ be orthogonal vectors of length $1 / \sqrt{2}$ in $\mathbb{R}^{n+1}$ which are orthogonal to both u and v. Then, the tangent space to V at (u, v) is spanned by the vectors

(7.3) $X_i = (e_i, 0) \; , \quad Y_i = (0, e_i) \; , \quad 1 \le i \le n-1 \; , \quad Z = (v, -u) \; .$

We first determine the shape operator A^ξ at (u, v). The curve

(7.4) $\gamma(t) = (\cos t \, u + \sin t \, e_i, v)$

has initial position $\gamma(0) = (u, v)$ and initial velocity vector $\gamma'(0) = (e_i, 0) = X_i$.

Along the curve γ, the normal field ξ is given by

$$\xi(t) = (-\cos t\, u - \sin t\, e_i, v)\,.$$

Thus, $\xi'(0) = (-e_i, 0) = -X_i$. Since $\xi'(0) = -A^\xi X_i$, we have $A^\xi X_i = X_i$, and each X_i is a principal vector with corresponding principal curvature 1. Similarly, the curve

(7.5) $\delta(t) = (u, \cos t\, v + \sin t\, e_i)$

has initial position (u, v) and initial velocity vector $(0, e_i) = Y_i$. Along the curve δ, we have

$$\xi(t) = (-u, \cos t\, v + \sin t\, e_i)\,.$$

Thus, $\xi'(0) = (0, e_i) = Y_i$, and $A^\xi Y_i = -Y_i$. Hence, each Y_i is a principal vector of A^ξ with corresponding principal curvature -1. Finally, the curve

(7.6) $\varepsilon(t) = (\cos t\, u + \sin t\, v, -\sin t\, u + \cos t\, v)$

has initial position (u, v) and initial velocity vector $(v, -u) = Z$. Along the curve ε, we have

$$\xi(t) = (-\cos t\, u - \sin t\, v, -\sin t\, u + \cos t\, v)\,.$$

Then $\xi'(0) = (-v, -u) = \eta$, which is normal to V. Therefore, the tangential component of $\xi'(0)$ is zero, and $A^\xi Z = 0$. Hence, we have found a basis of principal vectors of A^ξ as follows:

(7.7) $A^\xi X_i = X_i, \quad A^\xi Y_i = -Y_i, \quad 1 \le i \le n-1, \quad A^\xi Z = 0\,.$

We next determine A^η. Along the curve γ in (7.4), we have

$$\eta(t) = (-v, -\cos t\, u - \sin t\, e_i)\,.$$

Thus, $\eta'(0) = (0, -e_i) = -Y_i$ and $A^{\eta}X_i = Y_i$. Similarly, $A^{\eta}Y_i = X_i$.
Finally, along the curve ε in (7.6), we have

$$\eta(t) = (\sin t\, u - \cos t\, v, -\cos t\, u - \sin t\, v) .$$

Then $\eta'(0) = (u, -v) = -\xi$, which is normal to V, so $A^{\eta}Z = 0$. In summary,
we have

(7.8) $A^{\eta}X_i = Y_i , \quad A^{\eta}Y_i = X_i , \quad 1 \leq i \leq n-1 , \quad A^{\eta}Z = 0 .$

Any unit normal ζ to V at (u, v) can be written in the form

$$\zeta = \cos \theta\, \xi + \sin \theta\, \eta , \quad 0 \leq \theta \leq 2\pi .$$

If we take

$$U_i = \cos \frac{\theta}{2} X_i + \sin \frac{\theta}{2} Y_i , \qquad W_i = -\sin \frac{\theta}{2} X_i + \cos \frac{\theta}{2} Y_i ,$$

then (7.7) and (7.8) imply that

$$A^{\zeta}U_i = U_i , \quad A^{\zeta}W_i = -W_i , \quad 1 \leq i \leq n-1 , \quad A^{\zeta}Z = 0 .$$

Thus, A^{ζ} also has 3 principal curvatures $-1, 0, 1$ with respective multiplicities
$n-1, 1, n-1$, and we have proven (7.1).

Before constructing the counterexamples to the conjecture, it is helpful
to consider the curvature surfaces of V. According to our formulation, the
curvature surfaces are defined on the unit normal bundle $B(V)$. As we noted in
Remark 5.3, there is an equivalent definition in which the curvature surfaces
are defined on V itself. We recall that formulation in the following paragraph.

Suppose that M is any submanifold of codimension greater than one in
the sphere S^m. A connected submanifold $S \subset M$ is called a curvature surface if
there is a parallel section $\eta : S \to B(M)$ such that for each $x \in S$, the tangent
space T_xS is equal to some eigenspace of the shape operator $A^{\eta(x)}$.

Reckziegel [2] showed that if a principal curvature κ has constant multiplicity μ on $B(M)$ and is constant along its curvature surfaces, then all of its curvature surfaces are open subsets of μ–dimensional spheres in S^m. Since our particular submanifold V is compact, all of the curvature surfaces of κ_1, κ_2, and κ_3 are spheres of the appropriate dimension in S^{2n+1}.

Let α and β be positive real numbers satisfying $\alpha^2 + \beta^2 = 1$, and let $T^{\alpha,\beta}$ be the linear transformation of $\mathbb{R}^{2n+2}$ defined by

$$T^{\alpha,\beta}(x, y) = \sqrt{2}\,(\alpha x, \beta y) \ .$$

Note that the image $W^{\alpha,\beta}$ of V under $T^{\alpha,\beta}$ is contained in the sphere S^{2n+1}. In fact,

$$W^{\alpha,\beta} = \{(u, v) \mid |u| = \alpha \ , \ |v| = \beta \ , \ u \cdot v = 0\} \ .$$

We will show that the Legendre submanifold induced by $W^{\alpha,\beta}$ is a proper Dupin submanifold with four principal curvatures, which is not Lie equivalent to an isoparametric hypersurface if $\alpha \neq 1/\sqrt{2}$, i.e., if $T^{\alpha,\beta} \neq I$.

We first show that the Legendre submanifold induced by $W^{\alpha,\beta}$ is a proper Dupin submanifold with four principal curvatures $\lambda_1 < \lambda_2 < \lambda_3 < \lambda_4$. Since $W^{\alpha,\beta}$ has codimension two, the principal curvature $\lambda_4 = \infty$ has multiplicity one and is constant along its lines of curvature. To complete the proof that $W^{\alpha,\beta}$ is proper Dupin, we establish a bijective correspondence between the other curvature surfaces of V and those of $W^{\alpha,\beta}$. Let S be any curvature surface of V. Since V is compact and proper Dupin, S is a μ–dimensional sphere, where μ is the multiplicity of the corresponding principal curvature. Along the curvature surface S, the corresponding curvature sphere Σ is constant. Note that Σ is a hypersphere obtained by intersecting S^{2n+1} with a hyperplane π, which is tangent to V along S. The image $T^{\alpha,\beta}(\pi)$ is a hyperplane, which is tangent to $W^{\alpha,\beta}$ along the μ–dimensional sphere $T(S)$. Since the hypersphere $T^{\alpha,\beta}(\pi) \cap S^{2n+1}$ is tangent to $W^{\alpha,\beta}$ along $T(S)$, it is a curvature sphere of $W^{\alpha,\beta}$ with multiplicity μ, and $T(S)$ is the corresponding curvature surface. Thus, we have a bijective correspondence between the curvature surfaces of V and those of $W^{\alpha,\beta}$, and the Dupin condition is clearly satisfied on $W^{\alpha,\beta}$. Therefore, $W^{\alpha,\beta}$ is a proper Dupin submanifold with four principal curvatures, including $\lambda_4 = \infty$.

We now show that the Legendre submanifold induced by $W^{\alpha,\beta}$ is not Lie equivalent to an isoparametric hypersurface with four principal curvatures by showing that the Lie curvature Ψ does not equal $1/2$ at some points of the unit normal bundle $B(W^{\alpha,\beta})$. These calculations are similar to those for computing the principal curvatures of V above. First, note that

$$W^{\alpha,\beta} \subset f^{-1}(0) \cap g^{-1}(0) \,,$$

where f and g are the real–valued functions defined on S^{2n+1} by

$$f(u, v) = -\,(\beta\,/\,2\alpha)\,u \cdot u + (\alpha\,/\,2\beta)\,v \cdot v \,, \quad g(u, v) = -\,u \cdot v \,.$$

Thus, the gradients,

$$\xi = (\,-\frac{\beta}{\alpha}\,u,\,\frac{\alpha}{\beta}\,v\,) \,, \quad \eta = (-v,\,-u) \,,$$

of f and g are two unit normal vector fields to $W^{\alpha,\beta}$. Since $n \geq 2$, we can find x and y in $\mathbb{R}^{n+1}$ such that

$$|x| = \alpha \,, \quad x \cdot u = 0 \,, \quad x \cdot v = 0 \,,$$
$$|y| = \beta \,, \quad y \cdot u = 0 \,, \quad y \cdot v = 0 \,.$$

We define three curves,

$$\gamma(t) = (\cos t\,u + \sin t\,x,\,v) \,, \quad \delta(t) = (u,\,\cos t\,v + \sin t\,y) \,,$$

$$\varepsilon(t) = (\cos t\,u + \frac{\alpha}{\beta}\sin t\,v,\,-\frac{\beta}{\alpha}\sin t\,u + \cos t\,v) \,.$$

It is straightforward to check that each of these curves lies on $W^{\alpha,\beta}$ and goes through the point (u, v) when $t = 0$. Along the curve γ, the normal vector ξ is given by

$$\xi(t) = \left(\frac{-\beta}{\alpha}(\cos t\, u + \sin t\, x),\ \frac{\alpha}{\beta} v\right).$$

Then,

$$\xi'(0) = \left(\frac{-\beta}{\alpha} x,\, 0\right) = \frac{-\beta}{\alpha}\gamma'(0).$$

Thus, $X = \gamma'(0)$ is a principal vector of A^ξ at (u, v) with corresponding principal curvature β/α. Similarly, $Y = \delta'(0)$ is a principal vector of A^ξ at the point (u, v) with corresponding principal curvature $-\alpha/\beta$. Finally, along the curve ε, we have

$$\xi(t) = \left(\frac{-\beta}{\alpha}\left(\cos t\, u + \frac{\alpha}{\beta}\sin t\, v\right),\ \frac{\alpha}{\beta}\left(\frac{-\beta}{\alpha}\sin t\, u + \cos t\, v\right)\right).$$

Then $\xi'(0) = (-v, -u) = \eta$, which is normal to $W^{\alpha,\beta}$ at (u, v). Thus, we have $A^\xi(Z) = 0$ for $Z = \varepsilon'(0)$, and Z is a principal vector with corresponding principal curvature zero. Thus, at the point $\xi(u, v)$ in $B(W^{\alpha,\beta})$, there are four principal curvatures:

$$\lambda_1 = -\alpha/\beta,\quad \lambda_2 = 0,\quad \lambda_3 = \beta/\alpha,\quad \lambda_4 = \infty.$$

At this point, the Lie curvature is

$$\Psi = \Phi = (\lambda_1 - \lambda_2)/(\lambda_1 - \lambda_3) = (-\alpha/\beta)/(-\alpha/\beta - \beta/\alpha) = \alpha^2.$$

Hence, if $\alpha^2 \neq 1/2$, then the Legendre submanifold induced by $W^{\alpha,\beta}$ is not Lie equivalent to an isoparametric hypersurface in S^{2n+1} with four principal curvatures. To obtain a compact proper Dupin hypersurface in S^{2n+1} with four principal curvatures which is not Lie equivalent to an isoparametric hypersurface, one simply takes a tube of constant radius over $W^{\alpha,\beta}$ in S^{2n+1}.

Next, we handle the construction of the counterexamples due to Miyaoka and Ozawa [1]. The key ingredient here is the Hopf fibration of S^7 over S^4. Let $\mathbb{R}^8 = \mathbb{H} \times \mathbb{H}$, where $\mathbb{H}$ is the division ring of quaternions. The Hopf fibering of the unit sphere S^7 in $\mathbb{R}^8$ over S^4 is given by

$$(7.9) \qquad h(u, v) = (2u\bar{v}, |u|^2 - |v|^2) , \quad u, v \in \mathbb{H} .$$

One can easily compute that the image of h lies in the unit sphere S^4 in the Euclidean space $\mathbb{R}^5 = \mathbb{H} \times \mathbb{R}$.

Before beginning the construction of Miyaoka and Ozawa, we recall some facts about the Hopf fibration. Suppose $(w, t) \in S^4$, with $t \neq 1$, i.e., (w, t) is not the point $(0, 1)$. We want to find the inverse image of (w, t) under h. Suppose that

$$(7.10) \qquad 2u\bar{v} = w , \qquad |u|^2 - |v|^2 = t .$$

Multiplying the first equation in (7.10) by v on the right, we obtain

$$(7.11) \qquad 2u|v|^2 = wv , \qquad 2u = \frac{w}{|v|} \frac{v}{|v|} .$$

Since $|u|^2 + |v|^2 = 1$, the second equation in (7.10) yields

$$(7.12) \qquad |v|^2 = (1 - t) / 2 .$$

If we write $z = v / |v|$, then $z \in S^3$, the unit sphere in $\mathbb{H} = \mathbb{R}^4$. Then (7.11) and (7.12) give

$$(7.13) \qquad u = wz / (2(1 - t))^{1/2}, \quad v = ((1 - t) / 2)^{1/2} z , \quad z \in S^3.$$

Thus, if U is the open set $S^4 - \{(0, 1)\}$, then $h^{-1}(U)$ is diffeomorphic to $U \times S^3$ by the formula (7.13). Of course, the second equation in (7.10) shows that $h^{-1}(0, 1)$ is just the 3-sphere in S^7 determined by the condition $v = 0$. We can find a similar local trivialization containing these points with $v = 0$ by beginning the process above with multiplication of (7.10) by $\bar{u}$ on the left, rather than by v on the right. As a consequence of this local triviality, if M is an embedded submanifold in S^4 which does not equal all of S^4, then $h^{-1}(M)$ is diffeomorphic to $M \times S^3$. Finally, recall that the Euclidean inner product on the space $\mathbb{R}^8 = \mathbb{H} \times \mathbb{H}$ is given by

(7.14) $(a, b) \cdot (u, v) = \mathrm{Re}\, (\bar{a}u + \bar{b}v)$.

The examples of Miyaoka and Ozawa all arise as inverse images under h of proper Dupin hypersurfaces in S^4. The proof that these examples are proper Dupin is accomplished by first showing that they are taut. Thus, we begin with the following.

Theorem 7.1: *Let M be a compact, connected submanifold of S^4. If M is taut in S^4, then $h^{-1}(M)$ is taut in S^7.*

Proof: Since both M and $h^{-1}(M)$ lie in spheres, tautness is equivalent to the condition that every nondegenerate linear height function in $\mathbb{R}^8$ has the minimum number of critical points. We write linear height functions in $\mathbb{R}^8$ in the form

(7.15) $f_{ab}(u, v) = \mathrm{Re}\, (au + bv) = (\bar{a}, \bar{b}) \cdot (u, v) , \quad (a, b) \in S^7.$

This is the height function in the direction $(\bar{a}, \bar{b})$. We want to determine when a point (u, v) is a critical point of f_{ab} . Without loss of generality, we may assume that (u, v) is in a local trivialization of the form (7.13) when making local calculations. Let $x = (w, t)$ be a point of $M \subset S^4$, and let $(x, z) = (w, t, z)$ be a point in the fiber $h^{-1}(x)$. The tangent space to $h^{-1}(M)$ at (x, z) can be decomposed as $T_x M \times T_z S^3$. We first locate the critical points of the restriction of f_{ab} to the fiber through (x, z). By (7.13) and (7.15), we have

(7.16) $f_{ab}(w, t, z) = \mathrm{Re}\, (\, awz \,/\, (2(1-t))^{1/2} + bz\, ((1-t)/2)^{1/2} \,)$
$= \mathrm{Re}\, (\alpha(w, t)\, z) \;= \alpha(w, t) \cdot \bar{z} ,$

where

$\alpha(w, t) = aw \,/\, (2(1-t))^{1/2} + b\, ((1-t)/2)^{1/2} .$

This defines the map α from S^4 to $\mathbb{H}$. If Z is any tangent vector to S^3 at z, we write Zf_{ab} for the derivative of f_{ab} in the direction $(0, Z)$. Then

(7.17) $$Zf_{ab} = \alpha(w, t) \cdot Z$$

at (x, z). Now, there are two cases to consider. First, if $\alpha(w, t) \neq 0$, then in order to have $Zf_{ab} = 0$ for all Z in $T_z S^3$, we must have

(7.18) $$\bar{z} = \pm \alpha(w, t) / |\alpha(w, t)| .$$

So, the restriction of f_{ab} to the fiber has exactly 2 critical points with corresponding critical values

(7.19) $$\pm |\alpha(w, t)| .$$

The second case is when $\alpha(w, t) = 0$. Then, the restriction of f_{ab} to the fiber is identically zero by (7.16). In both cases, the function

$$g_{ab}(w, t) = |\alpha(w, t)|^2$$

satisfies the equation

$$g_{ab}(w, t) = f_{ab}^2(w, t, z)$$

at the critical point. The key in relating this fact to information about the submanifold M is to note that

(7.20)
$$\begin{aligned}
g_{ab}(w, t) = |\alpha(w, t)|^2 &= \frac{1}{2} \operatorname{Re} \{2a\bar{b}w + (|a|^2 - |b|^2) t\} + \frac{1}{2}(|a|^2 + |b|^2) \\
&= \frac{1}{2} + \frac{1}{2} ((w, t) \cdot (2a\bar{b}, |a|^2 - |b|^2)) \\
&= \frac{1}{2} + \frac{1}{2} \ell_{ab}(w, t) ,
\end{aligned}$$

where ℓ_{ab} is the linear height function in $\mathbb{R}^5$ in the direction

$$(2a\bar{b}, |a|^2 - |b|^2) = h(\bar{a}, \bar{b}) .$$

This shows that $g_{ab}(w, t) = 0$ if and only if $(w, t) = -h(\bar{a}, \bar{b})$. Thus, if $-h(\bar{a}, \bar{b})$

is not in M, the restriction of f_{ab} to each fiber has exactly two critical points of the form (x, z), with z as in (7.18). For $X \in T_xM$, we write Xf_{ab} for the derivative of f_{ab} in the direction $(X, 0)$. At the two critical points, we have

$$(7.21) \qquad\qquad Xf_{ab} = d\alpha(X) \cdot \bar{z} ,$$

$$(7.22) \quad Xg_{ab} = 2 \, d\alpha(X) \cdot \alpha(x) = \pm 2 \, |\alpha(x)| \, (d\alpha(X) \cdot \bar{z}) = \pm 2 \, |\alpha(x)| \, Xf_{ab} .$$

Thus, (x, z) is a critical point of f_{ab} if and only if x is a critical point of g_{ab}. By (7.20), this happens precisely when x is a critical point of ℓ_{ab}. We conclude that if $-h(\bar{a}, \bar{b})$ is not in M, then f_{ab} has two critical points for every critical point of ℓ_{ab} on M. The set of points (a, b) in S^7 such that $-h(\bar{a}, \bar{b})$ is in M has measure zero. If (a, b) is not in this set, then f_{ab} has twice as many critical points as the height function ℓ_{ab} on M. Since M is taut, every nondegenerate ℓ_{ab} has $\beta(M;\mathbb{F})$ critical points on M for some field $\mathbb{F}$. Thus, except for a set of measure zero, every height function f_{ab} has $2 \beta(M;\mathbb{F})$ critical points on $h^{-1}(M)$. Since $h^{-1}(M)$ is diffeomorphic to $M \times S^3$, we have

$$\beta(h^{-1}(M);\mathbb{F}) = \beta(M \times S^3;\mathbb{F}) = 2 \, \beta(M;\mathbb{F}) .$$

Thus, $h^{-1}(M)$ is taut in S^7.   □

We next use Theorem 7.1 to show that the inverse image under h of a compact proper Dupin submanifold in S^4 is proper Dupin. Recall that a submanifold M of codimension greater than one is proper Dupin if the Legendre submanifold induced by M is proper Dupin. Pinkall [5] proved that a taut submanifold is always Dupin, but it may not be proper Dupin, i.e., the number of distinct principal curvatures may not be constant on the unit normal bundle $B(M)$. Ozawa [1] proved that a taut submanifold $M \subset S^n$ is proper Dupin if and only if every connected component of a critical set of a linear height function on M is a point or is homeomorphic to a sphere of some dimension k. (See also Hebda [1].) This result is a key fact in the proof of the following theorem.

Theorem 7.2: *Let M be a compact, connected proper Dupin submanifold embedded in S^4. Then $h^{-1}(M)$ is a proper Dupin submanifold in S^7.*

Proof: As we have noted before, Thorbergsson [1] proved that a compact proper Dupin hypersurface embedded in S^n is taut, and Pinkall [5] extended this result to the case where M has codimension greater than one and the number of distinct principal curvatures is constant on $B(M)$. Thus, our M is taut in S^4, and therefore, $h^{-1}(M)$ is taut in S^7 by Theorem 7.1. To complete the proof of the theorem, we need to show that each connected component of a critical set of a height function f_{ab} on $h^{-1}(M)$ is a point or a sphere.

We use the same notation as in the proof of Theorem 7.1. Now, suppose that (x, z) is a critical point of f_{ab}. For $X \in T_x M$, we compute from (7.16) that

$$(7.23) \qquad X f_{ab} = d\alpha(X) \cdot \bar{z} .$$

From (7.22), we see that $X g_{ab}$ also equals zero, and the argument again splits into two cases, depending on whether or not $g_{ab}(x)$ is zero. If $g_{ab}(x)$ is non–zero, then there are two critical points of f_{ab} on the fiber $h^{-1}(x)$. Thus, a component in $h^{-1}(M)$ of the critical set of f_{ab} through (x, z) is homeomorphic to the corresponding component of the critical set containing x of the linear function ℓ_{ab} on M. Since M is proper Dupin, such a component is a point or a sphere.

The second case is when $g_{ab}(x) = f_{ab}^2(x, z) = 0$. As we have seen, this happens only if $x = -h(\bar{a}, \bar{b})$. In this case, x is an isolated absolute minimum of the height function ℓ_{ab}. Thus, the corresponding component of the critical set of f_{ab} through (x, z) lies in the fiber $h^{-1}(x)$, which is diffeomorphic to S^3. From (7.23), we see that this component of the critical set consists of those points (x, y) in the fiber such that $\bar{y}$ is orthogonal to $d\alpha(X)$, for all $X \in T_x M$. We know that

$$(7.24) \qquad g_{ab}(x) = \frac{1}{2} + \frac{1}{2} \ell_{ab}(x) ,$$

and x is an isolated critical point of ℓ_{ab} on M. The tautness of M and the results of Ozawa [1] imply that x is a nondegenerate critical point of ℓ_{ab}, since the component of the critical set of a height function containing a degenerate critical point must be a sphere of dimension greater than zero. By (7.24), x is also a nondegenerate critical point of g_{ab}, and so the Hessian $H(X, Y)$ of g_{ab} is nondegenerate at x. Since $\alpha(x) = 0$, we compute that for X and Y in T_xM,

$$H(X, Y) = 2\, d\alpha(X) \cdot d\alpha(Y).$$

Hence, $d\alpha$ is nondegenerate at x, and the rank of $d\alpha$ is the dimension of M. From this, it follows that the component of the critical set of f_{ab} through (x, z) is a sphere in $h^{-1}(x)$ of dimension $(3 - \dim M)$. Therefore, we have shown that every component of a critical set of a linear height function f_{ab} on $h^{-1}(M)$ is homeomorphic to a point or a sphere. Thus, $h^{-1}(M)$ is proper Dupin.   □

Next, we relate the principal curvatures of $h^{-1}(M)$ to those of M.

Theorem 7.3: *Let M be a compact, connected proper Dupin hypersurface embedded in S^4 with g principal curvatures. Then, the proper Dupin hypersurface $h^{-1}(M)$ in S^7 has $2g$ principal curvatures. Each principal curvature $\lambda = \cot\theta$, $0 < \theta < \pi$, of M at a point x yields two principal curvatures of $h^{-1}(M)$ at points in $h^{-1}(x)$ with values*

$$\lambda^+ = \cot(\theta/2), \qquad \lambda^- = \cot((\theta+\pi)/2).$$

Proof: A principal curvature $\lambda = \cot\theta$ of a hypersurface M at x corresponds to a focal point at oriented distance θ along the normal geodesic to M at x. (See, for example, Cecil–Ryan [7, p.127].) A point (x, z) in $h^{-1}(M)$ is a critical point of f_{ab} if and only if $(\bar{a}, \bar{b})$ lies along the normal geodesic to $h^{-1}(M)$ at (x, z). This critical point is degenerate if and only if $(\bar{a}, \bar{b})$ is a focal point of $h^{-1}(M)$ at x. Note further that (x, z) is a degenerate critical point of f_{ab} if and only if x is a degenerate critical point of ℓ_{ab}. This is true, since both embeddings are taut, and the dimensions of the components of the critical sets agree by Theorem 7.2. The latter claim holds even when $x = -h(\bar{a}, \bar{b})$, since the fact that M has dimension 3 implies that the critical point (x, z) of f_{ab} is isolated.

Thus, $(\bar{a}, \bar{b})$ is a focal point of $h^{-1}(M)$ if and only if $h(\bar{a}, \bar{b})$ is a focal point of M. Suppose now that $(\bar{a}, \bar{b})$ lies along the normal geodesic to $h^{-1}(M)$ at (x, z) and that $f_{ab}(x, z) = \cos \varphi$. Then, by (7.20),

$$g_{ab}(x) = \frac{1}{2} + \frac{1}{2}\,\ell_{ab}(x) = \frac{1}{2} + \frac{1}{2}\cos \theta \ ,$$

where θ is the distance from $h(\bar{a}, \bar{b})$ to x. Since (x, z) is a critical point of f_{ab}, we have $g_{ab}(x) = f_{ab}^{2}(x, z)$. Thus,

$$\frac{1}{2} + \frac{1}{2}\cos \theta = \cos^{2}\varphi = \frac{1}{2} + \frac{1}{2}\cos 2\varphi \ ,$$

and so $\cos \theta = \cos 2\varphi$. This means that under the map h, the normal geodesic to $h^{-1}(M)$ at (x, z) double covers the normal geodesic to M at x, since the points corresponding to the values $\varphi = \theta/2$ and $\varphi = (\theta+\pi)/2$ are mapped to the same point by h. In particular, a focal point corresponding to a principal curvature $\lambda = \cot \theta$ on the normal geodesic to M gives rise to two focal points on the normal geodesic to $h^{-1}(M)$, with corresponding principal curvatures

$$\lambda^{+} = \cot (\theta/2) , \quad \lambda^{-} = \cot ((\theta+\pi)/2) . \qquad \Box$$

We now construct the counterexamples to the conjecture, due to Miyaoka and Ozawa. Recall that a compact proper Dupin hypersurface M in S^{4} with two principal curvatures must be a cyclide of Dupin, but a conformal non–isometric image of an isoparametric cyclide does not have constant principal curvatures. Similarly, a compact proper Dupin hypersurface in S^{4} with three principal curvatures must be Lie equivalent to an isoparametric hypersurface, but it need not have constant principal curvatures itself.

Corollary 7.4: *Let M be a non-isoparametric compact, connected proper Dupin hypersurface in S^{4} with g principal curvatures, where $g = 2$ or 3. Then $h^{-1}(M)$ is a compact, connected proper Dupin hypersurface in S^{7} with $2g$ principal curvatures, which is not Lie equivalent to an isoparametric hypersurface in S^{7}.*

Proof: Suppose that $\lambda = \cot\theta$ and $\mu = \cot\alpha$ are two distinct non–constant principal curvature functions on M. Let

$$\lambda^+ = \cot(\theta/2)\,, \quad \lambda^- = \cot((\theta+\pi)/2)\,,$$
$$\mu^+ = \cot(\alpha/2)\,, \quad \mu^- = \cot((\alpha+\pi)/2)\,,$$

be the four distinct principal curvature functions on $h^{-1}(M)$ induced from λ and μ. Then the Lie curvature,

$$\Psi = (\lambda^+ - \lambda^-)(\mu^+ - \mu^-) \,/\, (\lambda^+ - \mu^-)(\mu^+ - \lambda^-) = 2\,/\,(1 + \cos(\theta - \alpha))\,,$$

is not constant on $h^{-1}(M)$, and therefore, $h^{-1}(M)$ is not Lie equivalent to an isoparametric hypersurface in S^7.   □

Certain parts of the construction of Miyaoka and Ozawa are also valid if ℍ is replaced by the Cayley numbers or even a more general Clifford algebra. See the paper of Miyaoka and Ozawa [1] for a discussion of this.

4

Dupin Submanifolds

In this chapter, we concentrate on local results which have been obtained using Lie sphere geometry. The main results are the classification of proper Dupin submanifolds with two principal curvatures (cyclides of Dupin) in Section 4.3 and the classification of proper Dupin hypersurfaces with three principal curvatures in $\mathbb{R}^4$ in Section 4.6. To obtain these classifications, we develop the method of moving Lie frames which can be used in the further study of Dupin submanifolds, or more generally, Legendre submanifolds.

4.1 Local Constructions

Pinkall [4] introduced four contructions for obtaining a Dupin hypersurface W in $\mathbb{R}^{n+m}$ from a Dupin hypersurface M in $\mathbb{R}^n$. These constructions involve building tubes, cylinders, cones and surfaces of revolution from M, and they will be discussed thoroughly in Section 4.2. Using these constructions, Pinkall was able to produce a proper Dupin hypersurface in Euclidean space with an arbitrary number of distinct principal curvatures, each with any given multiplicity (see Theorem 1.1 below). In general, these proper Dupin hypersurfaces cannot be extended to compact Dupin hypersurfaces without losing the property that the number of distinct principal curvatures is constant (see Section 4.2). For now, we give a proof of Pinkall's theorem without attempting to compactify the hypersurfaces constructed.

Theorem 1.1: *Given positive integers* $v_1,...,v_g$ *with* $v_1 + ... + v_g = n-1$, *there exists a proper Dupin hypersurface in* $\mathbb{R}^n$ *with* g *distinct principal curvatures having respective multiplicities* $v_1,...,v_g$.

Proof: The proof is by an inductive construction, which will be clear once the first few examples are done. To begin, note that a usual torus of revolution in $\mathbb{R}^3$ is a Dupin hypersurface with two distinct principal curvatures. To construct a proper Dupin hypersurface M^3 in $\mathbb{R}^4$ with three principal curvatures, each of multiplicity one, begin with an open subset U of a torus of revolution in $\mathbb{R}^3$ on which neither principal curvature vanishes. Take M^3 to be the cylinder $U \times \mathbb{R}$ in $\mathbb{R}^3 \times \mathbb{R} = \mathbb{R}^4$. Then M^3 has three distinct principal curvatures at each point, one of which is zero. These are clearly constant along their corresponding 1–dimensional curvature surfaces (lines of curvature).

To get a proper Dupin hypersurface in $\mathbb{R}^5$ with three principal curvatures having respective multiplicities $v_1 = v_2 = 1$, $v_3 = 2$, one simply takes $U \times \mathbb{R}^2$ in $\mathbb{R}^3 \times \mathbb{R}^2$. To obtain a proper Dupin hypersurface M^4 in $\mathbb{R}^5$ with four principal curvatures, first invert the hypersurface M^3 in a 3–sphere in $\mathbb{R}^4$, chosen so that the image of M^3 contains an open set W^3 on which no principal curvature vanishes. The hypersurface W^3 is Dupin, since the Dupin property is preserved by Lie transformations (Theorem 4.3 of Chapter 3). Now, take M^4 to be the cylinder $W^3 \times \mathbb{R}$ in $\mathbb{R}^4 \times \mathbb{R}$. □

We now turn to a full discussion of Pinkall's constructions in the setting of Lie sphere geometry.

4.2 Reducible Dupin Submanifolds

In this section, we study the standard constructions, introduced by Pinkall [4], for obtaining a proper Dupin submanifold μ with g distinct curvature spheres from a lower dimensional proper Dupin submanifold λ with $g-1$ curvature spheres. Pinkall only described these constructions locally, i.e., he began with a hypersurface M^{n-1} embedded in $\mathbb{R}^n$. Here, we formulate the constructions in the context of Lie sphere geometry. In each construction, we imitate the case where the Euclidean projection of λ is an immersion, but we do not assume this. We then determine the curvature spheres of μ and their multiplicities.

Pinkall listed four constructions which involve building tubes, cylinders, cones and surfaces of revolution over a Dupin submanifold. However, his Theorem 4 [4, p.438] showed that the cone construction is redundant, since it is Lie equivalent to a tube. This will be explained further in Remark 2.10. For this reason, we will only study three standard constructions: tubes, cylinders and surfaces of revolution.

A Dupin submanifold obtained from a lower dimensional Dupin submanifold via one of the standard constructions is said to be *reducible*. More generally, a Dupin submanifold which is locally Lie equivalent to such a Dupin submanifold is called reducible.

We first set some notation common to all three constructions. Let $\{e_1,...,e_{n+m+3}\}$ be the standard orthonormal basis for $\mathbb{R}_2^{n+m+3}$, with e_1 and e_{n+m+3} timelike. Let $\mathbb{P}^{n+m+2}$ be the projective space determined by $\mathbb{R}_2^{n+m+3}$, with corresponding Lie quadric Q^{n+m+1}. Let

$$\mathbb{R}_2^{n+3} = \text{Span }\{e_1,...,e_{n+2},e_{n+m+3}\} ,$$

and let $\mathbb{P}^{n+2}$ and Q^{n+1} be the corresponding projective space and quadric, respectively. Let Λ^{2n-1} and $\Lambda^{2(n+m)-1}$ be the spaces of projective lines on Q^{n+1} and Q^{n+m+1}, respectively. Finally, let $u_k = e_{k+2}$, $1 \leq k \leq n+m$, and

$$\mathbb{R}^n = \text{Span }\{e_3,...,e_{n+2}\} = \text{Span }\{u_1,...,u_n\} ,$$

$$\mathbb{R}^{n+m} = \text{Span }\{e_3,...,e_{n+m+2}\} = \text{Span }\{u_1,...,u_{n+m}\} .$$

A. *Tubes*

We will construct a Legendre submanifold which corresponds to building a tube in $\mathbb{R}^{n+m}$ around an $(n-1)$–dimensional submanifold of $\mathbb{R}^n$. This can be done for any Legendre submanifold, although we will assume that λ is a proper Dupin submanifold. We work with Euclidean projections of the Legendre submanifolds here, but one could just as well use spherical projections and construct a tube using the spherical metric (see Remark 2.10).

Let $\lambda : M^{n-1} \to \Lambda^{2n-1}$ be a proper Dupin submanifold with g distinct curvature spheres whose locus of point spheres does not contain the improper point $[e_1 - e_2]$. Then, the point sphere map $[k_1]$ and the hyperplane map $[k_2]$ for λ can be written as follows:

(2.1) $k_1 = (1 + f \cdot f, 1 - f \cdot f, 2f, 0) / 2$, $k_2 = (f \cdot \xi, -f \cdot \xi, \xi, 1)$.

These equations define the Euclidean projection f and Euclidean field of unit normals ξ for λ. As the calculations to follow will show, in order to construct a proper Dupin submanifold, we need to assume that f has constant rank. We will distinguish the case where f is an immersion from the case where f has lower rank. First, we assume that f is an immersion. The domain of the Legendre submanifold μ, corresponding to a tube of radius ε over f, is the unit normal bundle B^{n+m-1} to $f(M^{n-1})$ in $\mathbb{R}^{n+m}$. The normal vector fields $\xi, u_{n+1}, ..., u_{n+m}$ are all parallel with respect to the normal connection of $f(M^{n-1})$ in $\mathbb{R}^{n+m}$. This enables us to define a global trivialization of B^{n+m-1} with the properties of the local trivialization used in Section 3.3. Specifically, let

(2.2) $S^m = \{(y_0, ..., y_m) \mid y_0^2 + ... + y_m^2 = 1\}$.

Then, the map $(x, y) \to (x, \eta \, (x, y))$ with

(2.3) $\eta \, (x, y) = y_0 \xi(x) + y_1 u_{n+1} + ... + y_m u_{n+m}$

is a diffeomorphism from $M^{n-1} \times S^m \to B^{n+m-1}$. We now define the map

(2.4) $\mu : M^{n-1} \times S^m \to \Lambda^{2(n+m)-1}$,

corresponding to the tube of radius ε around $f(M^{n-1})$ in $\mathbb{R}^{n+m}$. This construction works for any real number ε. In particular, the case $\varepsilon = 0$ is just the Legendre submanifold induced by the immersion $f(M^{n-1})$ as a submanifold of codimension $m+1$ in $\mathbb{R}^{n+m}$. The map μ is defined by its Euclidean projection F and its Euclidean field of unit normals η (2.3), both of which are maps from $M^{n-1} \times S^m$ into $\mathbb{R}^{n+m}$. For $x \in M^{n-1}$ and $y = (y_0, ..., y_m) \in S^m$, we define the map μ by the formula $\mu \, (x, y) = [K_1(x, y), K_2(x, y)]$, where

(2.5) $K_1 = (1 + F \cdot F, 1 - F \cdot F, 2F, 0) / 2$, $K_2 = (F \cdot \eta, -F \cdot \eta, \eta, 1)$,

(2.6) $F(x, y) = f(x) + \varepsilon \, (y_0 \xi(x) + y_1 u_{n+1} + ... + y_m u_{n+m})$, and η is given by (2.3). To see that μ is a Legendre submanifold, we must check the Legendre conditions (1) $-$ (3) of Theorem 2.3 in Chapter 3. The Legendre condition (1) is easily verified. To check conditions (2) and (3) and find the curvature spheres, we must compute the differentials of K_1 and K_2 . We decompose the tangent space to $M^{n-1} \times S^m$ at the point $p = (x, y)$ as

$$(2.7) \qquad T_p(M^{n-1} \times S^m) = T_x M^{n-1} \times T_y S^m,$$

and denote a typical tangent vector by (X, Y). For K_1 and K_2 as in (2.5), it is easy to check that for real numbers r and s, at least one of which is not zero,

$$(2.8) \qquad d(rK_1 + sK_2)(X, Y) \in [K_1, K_2] \quad \Leftrightarrow \quad d(rF + s\eta)(X, Y) = 0 .$$

The tangent space (2.7) has a basis consisting of vectors of the form $(X, 0)$ and $(0, Y)$, where $Y = (Y_0, Y_1, ..., Y_m) \in T_y S^m$. From (2.3) and (2.6), we compute

$$(2.9) \qquad dF(X, 0) = df(X) + \varepsilon\, y_0\, d\xi(X) ,$$
$$(2.10) \qquad d\eta\, (X, 0) = y_0\, d\xi(X) ,$$
$$(2.11) \qquad dF(0, Y) = \varepsilon\, (Y_0 \xi(x) + Y_1 u_{n+1} + ... + Y_m u_{n+m}) ,$$
$$(2.12) \qquad d\eta\, (0, Y) = Y_0 \xi(x) + Y_1 u_{n+1} + ... + Y_m u_{n+m} .$$

The Legendre contact condition (3) reduces to $dF \cdot \eta = 0$ for K_1, K_2 as in (2.5). This can now be checked directly from (2.3), (2.9) and (2.11), using the fact that $df(X)$ and $d\xi(X)$ are both orthogonal to $\xi(x)$ and that $y_0 Y_0 + ... + y_m Y_m = 0$.

Next we locate the curvature spheres of μ. The fact that condition (2) is satisfied follows from these calculations. From (2.11) and (2.12), we see that

$$d(F - \varepsilon\eta)(0, Y) = 0 ,$$

for every $Y \in T_y S^m$. Thus, $[K_1 - \varepsilon K_2]$ is a curvature sphere of multiplicity at least m at each point of $M^{n-1} \times S^m$. From formulas (3.4) and (3.6) of Chapter 1, we see that $[K_1 - \varepsilon K_2]$ represents an oriented hypersphere with center $f(x)$ and radius $-\varepsilon$ (the minus sign is due to the outward normal). This is the new family of curvature spheres which results from this construction. Note that if there were a non–zero vector $X \in T_x M^{n-1}$ such that $df(X) = 0$, then we would have $d(F - \varepsilon\eta)(X, 0) = 0$ also, and the curvature sphere $[K_1 - \varepsilon K_2]$ would have multiplicity $m + \nu$ at (x, y), where ν is the nullity of df at x. This shows that f must have constant rank for this construction to yield a proper Dupin submanifold. Since we have assumed that f is an immersion, $[K_1 - \varepsilon K_2]$ has constant multiplicity m. The curvature surfaces of $[K_1 - \varepsilon K_2]$ are of the form $\{x\} \times S^m$, for x in M^{n-1}, and $[K_1 - \varepsilon K_2]$ is constant along these curvature surfaces. From (3.4) of Chapter 1, the analytic condition for these curvature spheres to have radius $-\varepsilon$ is

(2.13) $< K_1 - \varepsilon K_2 \,,\, \varepsilon\, e_1 - \varepsilon\, e_2 + e_{n+m+3} > = 0 \,.$

The centers of these spheres all lie in $\mathbb{R}^n$, i.e.,

$$< K_1 - \varepsilon K_2 \,,\, u_{n+i} > = < K_1 - \varepsilon K_2 \,,\, e_{n+2+i} > = 0 \,, \quad 1 \le i \le m \,.$$

Thus, the image of the map $[K_1 - \varepsilon K_2]$ is contained in the $(n+1)$–dimensional linear subspace $E \subset \mathbb{P}^{n+m+2}$ whose orthogonal complement is the space

(2.14) $E^{\perp} = \text{Span } \{e_{n+3}, ..., e_{n+m+2}, \varepsilon\, e_1 - \varepsilon\, e_2 + e_{n+m+3}\} \,,$

which has signature $(m, 1)$.

The computation to determine the other curvature spheres splits into two cases, depending on whether the coordinate y_0 of y is zero.

Case 1: $y_0 \ne 0$. This is the case when the vector η in (2.3) is not orthogonal to $\mathbb{R}^n$. Then, from (2.9) – (2.10) we have for any real numbers r and s,

$$d(rF + s\eta)(X, 0) = r\, df(X) + y_0\, (r\varepsilon + s)\, d\xi(X) \,.$$

Since $y_0 \ne 0$, we can obtain all possible linear combinations of $df(X)$ and $d\xi(X)$ by appropriate choices of r and s. Thus, there exist numbers r and s such that

$$d(rF + s\eta)(X, 0) = 0 \,,$$

precisely when $df(X)$ and $d\xi(X)$ are linearly dependent, i.e., X is a principal vector of the original Legendre submanifold λ. Hence, the other curvature spheres, with their respective multiplicities, correspond to the curvature spheres of λ at x. Since λ is Dupin, these curvature spheres of μ are constant along their curvature surfaces, which have the form $S \times \{y\}$, where S is a curvature surface of λ through x. The number $\gamma(x, y)$ of curvature spheres at the point (x, y) equals $g+1$, where g is the number of curvature spheres of λ.

Case 2: $y_0 = 0$. This is the case when η is orthogonal to $\mathbb{R}^n$. At these points, we have from (2.9) $-$ (2.10) that

$$(2.15) \qquad dF(X, 0) = df(X) , \quad d\eta(X, 0) = 0 .$$

Formula (2.15) implies that $[K_2]$ is a curvature sphere of multiplicity $n-1$ at the point (x, y). Hence, at (x, y) there are only two distinct curvature spheres, $[K_2]$ and $[K_1 - \varepsilon K_2]$, having respective multiplicities $n-1$ and m . These curvature spheres are constant along their curvature surfaces, which have the form $M^{n-1} \times \{y\}$ and $\{x\} \times S^m$, respectively. This holds regardless of the original Legendre submanifold λ. In terms of Euclidean geometry, the points satisfying $y_0 = 0$ correspond to points at a distance ε away from the original space $\mathbb{R}^n$. (See Example 4.6 of Chapter 3 of a tube over a torus $T^2 \subset \mathbb{R}^3 \subset \mathbb{R}^4$.) The set of points where $y_0 = 0$ is diffeomorphic to $M^{n-1} \times S^{m-1}$. We summarize our results for the tube construction in the following proposition.

Proposition 2.1: *Suppose that* $\lambda : M^{n-1} \to \Lambda^{2n-1}$ *is a proper Dupin submanifold with g distinct curvature spheres such that the Euclidean projection f is an immersion of* M^{n-1} *into* $\mathbb{R}^n \subset \mathbb{R}^{n+m}$. *Then, the tube construction yields a Dupin submanifold* μ *defined on the unit normal bundle* B^{n+m-1} *to* $f(M^{n-1})$ *in* $\mathbb{R}^{n+m}$. *The number* $\gamma(x, \eta)$ *of distinct curvature spheres of* μ *at a point* $(x, \eta) \in B^{n+m-1}$ *is as follows:*

(a) $\gamma(x, \eta) = 2$, *if* η *is orthogonal to* $\mathbb{R}^n$ *in* $\mathbb{R}^{n+m}$.

(b) $\gamma(x, \eta) = g+1$, *otherwise.*

Remark 2.2: In the case $\varepsilon = 0$, μ is the Legendre submanifold induced by the immersion $f(M^{n-1})$ as a submanifold of codimension $m+1$ in $\mathbb{R}^{n+m}$. Proposition 5.2 of Chapter 3 describes the curvature spheres of μ. The point sphere map $[K_1]$ of (2.5) is a curvature sphere of multiplicity m, which lies in the $(n+1)$–dimensional subspace E with orthogonal complement $E^\perp$ in (2.14) with $\varepsilon = 0$. The tubes of radius $\varepsilon \neq 0$ over $f(M^{n-1})$ are parallel submanifolds of μ.

We now assume that the Euclidean projection f of λ has constant rank less than $n-1$. Then, λ is the Legendre submanifold induced by an immersed submanifold $\varphi : V \to \mathbb{R}^n$ of codimension $v+1$, and the domain of λ is the unit normal bundle B^{n-1} of $\varphi(V)$ in $\mathbb{R}^n$. As in Remark 2.2, we first consider the Legendre submanifold μ induced by the submanifold $\varphi(V)$ of codimension $m+v+1$ in $\mathbb{R}^{n+m}$. The number of distinct curvature spheres of μ at a point (x, η) in B^{n+m-1} is determined by Proposition 5.2 of Chapter 3. We decompose the unit vector η as follows:

$$\eta = \cos \theta \, \xi + \sin \theta \, u \, , \quad 0 \leq \theta \leq \pi/2 \, ,$$

where ξ is a unit vector in $\mathbb{R}^n$ normal to $\varphi(V)$ at $\varphi(x)$, and u is orthogonal to $\mathbb{R}^n$ in $\mathbb{R}^{n+m}$. Since the shape operator $A^u = 0$, we have $A^\eta = \cos \theta \, A^\xi$. If $\cos \theta$ is non–zero, then A^η has the same number of distinct principal curvatures as A^ξ. Thus, the number of distinct curvature spheres of μ at (x, η) is the same as the number of distinct curvature spheres of λ at (x, ξ), since the point sphere map is a curvature sphere in both cases. On the other hand, if $\cos \theta = 0$, then we have $A^\eta = 0$, and the number of distinct curvature spheres of μ at (x, η) is two. This is similar to case (a) in Proposition 2.1.

Using the local trivialization of B^{n+m-1} given in Section 3.3, it is easy to check that μ is Dupin, since we are assuming that λ is Dupin. Furthermore, the point sphere map $[K_1]$ is a curvature sphere of multiplicity $m+v$, and it lies in an $(n+1)$–dimensional subspace E of $\mathbb{P}^{n+m+2}$ whose orthogonal complement is the space $E^\perp$ in (2.14) with $\varepsilon = 0$.

The Legendre submanifold corresponding to a tube of radius $\varepsilon \neq 0$ over $\varphi(V)$ in $\mathbb{R}^{n+m}$ is a parallel submanifold to μ. Thus, it is also Dupin, and it has the same number of curvature spheres at each point as μ. We summarize these results in the following proposition.

Proposition 2.3: *Suppose that* $\lambda : B^{n-1} \to \Lambda^{2n-1}$ *is a proper Dupin submanifold with g distinct curvature spheres induced by an immersed submanifold $\varphi(V)$ of codimension $\nu+1$ in* $\mathbb{R}^n \subset \mathbb{R}^{n+m}$. *Then, the tube construction yields a Dupin submanifold μ defined on the unit normal bundle* B^{n+m-1} *to $\varphi(V)$ in* $\mathbb{R}^{n+m}$. *The number $\gamma(x, \eta)$ of distinct curvature spheres of μ at a point (x, η) in* B^{n+m-1} *is as follows:*

(a) $\gamma(x, \eta) = 2$, *if η is orthogonal to* $\mathbb{R}^n$ *in* $\mathbb{R}^{n+m}$.

(b) $\gamma(x, \eta) = g$, *otherwise.*

Remark 2.4: The original purpose of Pinkall's constructions was to increase the number of distinct curvature spheres by one, as in Proposition 2.1. However, as Proposition 2.3 shows, this does not happen when λ is induced by a submanifold $\varphi(V)$ of codimension greater than one in $\mathbb{R}^n$. Still, we consider the Dupin submanifold μ in Proposition 2.3 to be reducible, since it is obtained from λ by one of the standard constructions.

B. *Cylinders*

As before, we begin with a proper Dupin submanifold $\lambda : M^{n-1} \to \Lambda^{2n-1}$ with g distinct curvature spheres at each point, and assume that the locus of point spheres does not include the improper point $[e_1 - e_2]$. We can write the point sphere map $[k_1]$ and hyperplane map $[k_2]$ in the form (2.1), and thereby define the Euclidean projection f and Euclidean field of unit normals ξ as maps from M^{n-1} to $\mathbb{R}^n$. Usually, one thinks of the cylinder built over f in $\mathbb{R}^{n+m} = \mathbb{R}^n \times \mathbb{R}^m$ to be the map from $M^{n-1} \times \mathbb{R}^m$ to $\mathbb{R}^{n+m}$ given by $(x, z) \to f(x) + z$. Here, we attempt to extend this to a map defined on $M^{n-1} \times S^m$ by working in the context of Lie geometry. This is accomplished by mapping all points in the set $M^{n-1} \times \{\infty\}$ to the improper point in Lie geometry. The Legendre immersion condition (2) can still be satisfied at points (x, ∞) because the normal vector varies as x varies. However, as the computations below show, the Legendre immersion condition (2) is only satisfied at points of the form (x, ∞) for which the map ξ has rank $n-1$ at x.

As in the tube construction, we consider S^m as in (2.2). We relate S^m to $\mathbb{R}^m$ via stereographic projection τ from $S^m - \{P\}$, where $P = (-1,0,...,0)$, to $\mathbb{R}^m$,

$$\tau\,(y_0,...,y_m) = \frac{1}{1+y_0}\,(y_1,...,y_m) = (z_1,...,z_m) = z\;.$$

We define the Legendre submanifold μ corresponding to the cylinder over f in $\mathbb{R}^{n+m}$ by giving its point map F and Euclidean field of unit normals η. For a cylinder defined on $M^{n-1} \times \mathbb{R}^m$, these should obviously be defined as follows:

(2.16) $F(x, z) = f(x) + z_1 u_{n+1} + ... + z_m u_{n+m}\;,$

(2.17) $\eta(x, z) = \xi(x)\;.$

We can now obtain the extension to $M^{n-1} \times \{P\}$ by writing the maps K_1 and K_2 induced from F and η in the usual way (2.5). First, note that

$$F\cdot F = f\cdot f + \sum_{i=1}^{m} z_i^2 = f\cdot f + \frac{1-y_0^2}{(1+y_0)^2} = f\cdot f + \frac{1-y_0}{1+y_0}\;.$$

We can multiply K_1 by $2(1 + y_0)$ and get

$$[K_1] = [(2 + (f\cdot f)(1+y_0)\,,\,2y_0 - (f\cdot f)(1+y_0)\,,$$

$$2(1+y_0)\,f + 2y_1 u_{n+1} + ... + 2y_m u_{n+m}\,,\,0)]\;.$$

When the values $y_0 = -1$ and $y_i = 0$, $1 \le i \le m$, are substituted into this formula, we get $[K_1] = [(1,-1,0,...,0)]$, the improper point, as desired.

Since the formula for $K_2(x, y)$ does not depend on the point $y \in S^m$, it does not need to be modified to incorporate the special point P. Specifically,

$$K_2(x, y) = (\,f(x)\cdot\xi(x),\,-f(x)\cdot\xi(x),\,\xi(x),\,1\,)\;.$$

As with the tube construction, it is easy to check that K_1 and K_2 satisfy the Legendre condition (1). To check conditions (2) and (3) and to locate the curvature spheres, we compute the differentials of K_1 and K_2. We first consider points (x, y) with $y_0 \neq -1$, that is, $y \neq P$. For these points, it is simpler to use formulas (2.16) – (2.17) defined on $M^{n-1} \times \mathbb{R}^m$. Thus, we identify a point $y \in S^m - \{P\}$ with the point $z = \tau(y)$ in $\mathbb{R}^m$, and identify $T_y S^m$ with $T_z \mathbb{R}^m$ via $d\tau$. We write a typical tangent vector as (X, Z), for X in $T_x M^{n-1}$ and Z in $T_z \mathbb{R}^m$. As before, the location of the curvature spheres is determined by the condition

$$d(rF + s\eta)\,(X, Z) = 0 \ .$$

If we write $Z = (Z_1,...,Z_m)$, then we can compute from (2.16) – (2.17) that

(2.18) $\qquad dF(X, 0) = df(X) \ , \qquad d\eta(X, 0) = d\xi(X) \ ,$

(2.19) $\qquad dF(0, Z) = Z_1 u_{n+1} + ... + Z_m u_{n+m} \ , \qquad d\eta(0, Z) = 0 \ .$

The Legendre condition (3), $dF \cdot \eta = 0$, is easily verified using (2.17)–(2.19) and the Legendre condition for λ, $df \cdot \xi = 0$. From (2.19), it follows immediately that $[K_2]$ is a curvature sphere of multiplicity at least m at each point of $M^{n-1} \times \mathbb{R}^m$. This is also true on $M^{n-1} \times S^m$, since $[K_2]$ does not depend on the point $y \in S^m$. At each point, $[K_2]$ corresponds to a hyperplane in $\mathbb{R}^{n+m}$ in oriented contact with the cylinder along one of its rulings. Note further that if there is a non–zero vector $X \in T_x M^{n-1}$ such that $d\xi(X) = 0$, then $d\eta\,(X, 0) = 0$ also, and the curvature sphere $[K_2]$ has multiplicity $m+\nu$, where ν is the nullity of $d\xi$ at x. Thus, $[K_2]$ does not have constant multiplicity unless ξ has constant rank on M^{n-1}.

Therefore, we now assume that ξ has constant rank on M^{n-1}. We first consider the case when the rank of ξ is $n-1$, i.e., ξ is an immersion on M^{n-1}. Then, $[K_2]$ has multiplicity m and is constant along its curvature surfaces, which have the form $\{x\} \times S^m$. Since $[K_2]$ represents a hyperplane, it satisfies the equation $< K_2, e_1 - e_2 > = 0$. Furthermore, the normal $\eta(x, y) = \xi(x)$ to the hyperplane lies in $\mathbb{R}^n$, and so

$$< K_2, u_{n+i} > = < K_2, e_{n+2+i} > = 0 \ , \qquad 1 \leq i \leq m \ .$$

Hence, the image of the map $[K_2]$ is contained in the $(n+1)$–dimensional linear subspace E in $\mathbb{P}^{n+2}$ whose orthogonal complement is the $(m+1)$–dimensional vector space,

$$(2.20) \qquad E^\perp = \text{Span } \{e_{n+3},...,e_{n+m+2}, e_1 - e_2\},$$

on which the scalar product $< , >$ has signature $(m, 0)$.

From (2.18), we see that the other curvature spheres correspond exactly to the curvature spheres of the original Dupin submanifold λ. These curvature spheres are also constant along their curvature surfaces, which are of the form $S \times \{y\}$, where S is a curvature surface of λ. Thus, μ has $g+1$ curvature spheres at the points (x, y) with $y \neq P$.

We now consider points in $M^{n-1} \times \{P\}$. Let x be an arbitrary point of M^{n-1}. We know that $dK_1(X, 0) = 0$ at (x, P) for all $X \in T_x M^{n-1}$, since $[K_1]$ is constant on the set $M^{n-1} \times \{P\}$. On the other hand, one can easily compute that $dK_1(0, Y)$ is not in $[K_1, K_2]$ for any Y. Thus, $[K_1]$ is a curvature sphere of multiplicity $n-1$ at each point (x, P). By differentiating the formula for K_2 and using the fact that $df \cdot \xi = 0$, we get

$$(2.21) \qquad dK_2(X, 0) = (\, f \cdot d\xi(X), -f \cdot d\xi(X), d\xi(X), 0 \,) \,.$$

Since we have assumed that ξ is an immersion, $dK_2(X, 0) \neq 0$ for $X \neq 0$. This shows that the Legendre condition (2) is satisfied and that there are exactly two distinct curvature spheres, $[K_1]$ with multiplicity $n-1$ and $[K_2]$ with multiplicity m at (x, P). These have respective curvature surfaces $M^{n-1} \times \{P\}$ and $\{x\} \times S^m$. The Dupin condition is clearly satisfied by these curvature surfaces, and thus μ is Dupin. We summarize these results as follows.

Proposition 2.5: *Suppose that $\lambda : M^{n-1} \to \Lambda^{2n-1}$ is a proper Dupin submanifold with g distinct curvature spheres such that the Euclidean field of unit normals ξ is an immersion. Then, the cylinder construction yields a Dupin submanifold μ defined on $M^{n-1} \times S^m$. The number $\gamma(x, y)$ of distinct curvature spheres of μ at a point $(x, y) \in M^{n-1} \times S^m$ is as follows:*

(a) $\gamma(x, y) = 2$, *if y is the pole P of the stereographic projection τ from S^m to $\mathbb{R}^m$.*

(b) $\gamma(x, y) = g + 1$, *otherwise.*

The cylinder construction also yields a Dupin submanifold defined on $M^{n-1} \times \mathbb{R}^m$ if ξ has constant rank $n-1-\nu$, $\nu \geq 1$. However, the construction does not extend to $M^{n-1} \times \{P\}$ because the Legendre immersion condition (2) is not satisfied at those points. Specifically, if $X \in T_x M^{n-1}$ is a non–zero vector such that $d\xi(X) = 0$, then $dK_1(X, 0)$ and $dK_2(X, 0)$ are both zero at the point (x, P). Furthermore, the number of distinct curvature spheres of the cylinder μ on $M^{n-1} \times \mathbb{R}^m$ is g, not $g+1$, since the hyperplane map is already a curvature sphere of λ. The curvature surfaces of $[K_2]$ are of the form $S \times S^m$, where S is a curvature surface of λ. Thus, we have:

Proposition 2.6: *Suppose that $\lambda : M^{n-1} \to \Lambda^{2n-1}$ is a proper Dupin submanifold with g distinct curvature spheres such that the Euclidean field of unit normals ξ has constant rank $n-1-\nu$, $\nu \geq 1$. Then, the cylinder construction yields a proper Dupin submanifold μ defined on $M^{n-1} \times \mathbb{R}^m$ with g distinct curvature spheres at each point.*

C. *Surfaces of revolution*

As before, we begin with a proper Dupin submanifold $\lambda : M^{n-1} \to \Lambda^{2n-1}$ with g distinct curvature spheres, and assume that the point sphere map does not contain the improper point. We write the point sphere map $[k_1]$ and hyperplane map $[k_2]$ in the form (2.1), thereby determining f and ξ. We want to construct the Legendre submanifold μ obtained by "revolving" f around an axis $\mathbb{R}^{n-1} \subset \mathbb{R}^n \subset \mathbb{R}^{n+m}$. We do not assume that f is an immersion, nor that the image of f is disjoint from the axis $\mathbb{R}^{n-1}$. For simplicity, we assume that the axis goes through the origin in $\mathbb{R}^n$ and that the standard orthonormal basis vectors $u_1,...,u_{n+m}$ have been chosen so that $\mathbb{R}^{n-1}$ is the span of $u_1,...,u_{n-1}$. We write the sphere S^m in the form

$$S^m = \{y = y_0 u_n + y_1 u_{n+1} + ... + y_m u_{n+m} \mid y_0^2 + ... + y_m^2 = 1\} .$$

We can define the new Legendre submanifold $\mu : M^{n-1} \times S^m \to \Lambda^{2(n+m)-1}$ by giving its Euclidean projection F and Euclidean field of unit normals η. First, we decompose the maps f and ξ into components along $\mathbb{R}^{n-1}$ and orthogonal to $\mathbb{R}^{n-1}$ and write

$$f(x) = \hat{f}(x) + f_n(x) , \quad \hat{f}(x) \in \mathbb{R}^{n-1},$$
$$\xi(x) = \hat{\xi}(x) + \xi_n(x) , \quad \hat{\xi}(x) \in \mathbb{R}^{n-1}.$$

For $x \in M^{n-1}, y \in S^m$, we let

(2.22) $F(x, y) = \hat{f}(x) + f_n(x)\, y ,$

(2.23) $\eta(x, y) = \hat{\xi}(x) + \xi_n(x)\, y .$

K_1 and K_2 are then defined by (2.5), as before. Again, it is easy to check that the Legendre condition (1) is satisfied. To check (2), (3) and locate the curvature spheres of μ, we compute dF and $d\eta$. We consider vectors X in $T_x M^{n-1}$ and $Y = Y_0 u_n + ... + Y_m u_{n+m}$ in $T_y S^m$, and compute

(2.24) $dF(X, 0) = d\hat{f}(X) + (Xf_n)y , \quad d\eta\, (X, 0) = d\hat{\xi}(X) + (X\xi_n)y ,$

(2.25) $dF(0, Y) = f_n(x)\, Y , \quad d\eta\, (0, Y) = \xi_n(x)\, Y .$

Note that when $y = u_n$, we have $dF(X, 0) = df(X)$ and $d\eta(X, 0) = d\xi(X)$. The Legendre condition (3), $dF \cdot \eta = 0$, follows easily from $df \cdot \xi = 0$ and $y \cdot Y = 0$. To check (2), first note that $dF(X, 0)$ and $d\eta(X, 0)$ are never simultaneously zero, since $df(X)$ and $d\xi(X)$ are never both zero for $X \neq 0$. Then, (2.24) and (2.25) imply that $dF(X, Y)$ and $d\eta(X, Y)$ are never both zero if $X \neq 0$. On the other hand, $dF(0, Y)$ and $d\eta(0, Y)$ vanish simultaneously at (x, y) for a non–zero $Y \in T_y S^m$ if and only if $f_n(x)$ and $\xi_n(x)$ are both zero, i.e., $f(x)$ and $\xi(x)$ both lie in $\mathbb{R}^{n-1}$. This means that the line through $f(x)$ in the direction $\xi(x)$ lies in $\mathbb{R}^{n-1}$. This line is the locus of centers of the spheres in $\mathbb{R}^n$ corresponding to the points on the line $\lambda(x)$ in Q^{n+1}. These spheres all meet $\mathbb{R}^{n-1}$ orthogonally, as does the one plane in the parabolic pencil determined by $\lambda(x)$.

Condition (2) does not hold at points (x, y) of this type. This is easy to understand geometrically. If $f(x)$ and $\xi(x)$ both lie in $\mathbb{R}^{n-1}$, then they are fixed by the group $SO(m+1)$ of isometries of $\mathbb{R}^{n+m}$ which keep $\mathbb{R}^{n-1}$ pointwise fixed. So, the contact element $(f(x), \xi(x))$ is also fixed by these rotations, and the map μ from $M^{n-1} \times S^m$ into the space of contact elements is not an immersion at such points.

We now find the curvature spheres of μ at a point (x, y) where the Legendre condition (2) holds. As before, the curvature spheres are determined by the condition

$$d(rF + s\eta)\,(X, Y) = 0 \ .$$

From (2.24), we see that $d(rF+s\eta)(X, 0) = 0$ if and only if $d(rf+s\xi)(X) = 0$. Thus, $(X, 0)$ is a principal vector for μ at (x, y) if and only if X is a principal vector for λ at x. The curvature sphere of μ with principal vector $(X, 0)$ corresponds to the curvature sphere of λ with principal vector X.

The new curvature sphere of μ is easily found from (2.25). For any Y in $T_y S^m$, and for any fixed x, we have

$$d(\xi_n(x)F - f_n(x)\eta)\,(0, Y) = 0 \ .$$

Hence, $[K] = [\xi_n(x)K_1 - f_n(x)K_2]$ is a curvature sphere of multiplicity at least m at (x, y). From (3.6) of Chapter 1, we see that if $\xi_n(x) \neq 0$, then $[K]$ represents the oriented sphere with center $\xi_n(x)\hat{f}(x) - f_n(x)\hat{\xi}(x)$ and signed radius $-f_n(x)/\xi_n(x)$. The center of $[K]$ is the unique point of intersection of the line through $f(x)$ in the direction $\xi(x)$ with the axis $\mathbb{R}^{n-1}$, and $[K]$ meets $\mathbb{R}^{n-1}$ orthogonally. If $\xi_n(x) = 0$, then $[K]$ represents the hyperplane through $f(x)$ with normal $\xi(x) = \hat{\xi}(x)$. This hyperplane also meets $\mathbb{R}^{n-1}$ orthogonally. Thus, in either case, $[K]$ is orthogonal to $u_{n+i} = e_{n+2+i}$, $0 \leq i \leq m$, and $[K]$ is contained in the $(n+1)$–dimensional linear subspace E of $\mathbb{P}^{n+m+2}$ whose orthogonal complement is the $(m+1)$–dimensional vector space

$$(2.26) \qquad\qquad E^\perp = \mathrm{Span}\ \{e_{n+2},...,e_{n+m+2}\}\ ,$$

which has signature $(m+1, 0)$.

There are two possibilities for the number of distinct curvature spheres of μ at (x, y). The first case is when $[K]$ is not equal to one of the curvature spheres induced from the curvature spheres of λ at x, i.e., when none of the curvature spheres on the line $\lambda(x)$ are orthogonal to $\mathbb{R}^{n-1}$. In that case, $[K]$ has multiplicity m, and its curvature surface through (x, y) is $\{x\} \times S^m$, along which $[K]$ is constant. The other curvature spheres of μ at (x, y) correspond exactly to the curvature spheres of λ at x. Since λ is Dupin, these curvature spheres of μ are also constant along their curvature surfaces, which have the form $S \times \{y\}$, where S is a curvature surface of λ through x. The number $\gamma(x, y)$ of distinct curvature spheres of μ at (x, y) is $g+1$. The second case is when $[K]$ is the curvature sphere induced from one of the curvature spheres $[k]$ of λ at x. Then, $[K]$ has multiplicity $m+\nu$, where ν is the multiplicity of $[k]$. The curvature surface of $[K]$ through (x, y) is $S \times S^m$, where S is the curvature surface of $[k]$ through x. The curvature sphere $[K]$ is clearly constant along this curvature surface. In this case, $\gamma(x, y) = g$. We summarize these results in the following proposition.

Proposition 2.7: *Suppose that $\lambda : M^{n-1} \to \Lambda^{2n-1}$ is a proper Dupin submanifold with g distinct curvature spheres. The surface of revolution construction yields a Dupin submanifold μ defined on all of $M^{n-1} \times S^m$, except those points where the spheres in the parabolic pencil determined by the line $\lambda(x)$ are all orthogonal to the axis $\mathbb{R}^{n-1}$. For (x, y) in the domain of μ, the number $\gamma(x, y)$ of distinct curvature spheres at (x, y) is as follows:*

(a) $\gamma(x, y) = g+1$, *if none of the curvature spheres of λ at x are orthogonal to the axis $\mathbb{R}^{n-1}$.*

(b) $\gamma(x, y) = g$, *otherwise.*

Thus, as with the other constructions, there are two cases in which this construction yields a proper Dupin submanifold; either no curvature sphere of λ is ever orthogonal to the axis or one of the curvature spheres of λ is always orthogonal to the axis.

We now find a Lie geometric criterion to determine when a Dupin submanifold is reducible to a lower dimensional Dupin submanifold. First, note that the umbilic case of a proper Dupin submanifold with one distinct curvature sphere is well known. These are all Lie equivalent to the Legendre submanifold induced from an open subset of a standard sphere S^{n-1} embedded in $\mathbb{R}^n$. A standard sphere is reducible to a point by any of the standard constructions. From now on, we only consider the case in which the number of distinct curvature spheres is greater than one.

In general, as we see from Propositions 2.1, 2.5 and 2.7, the application of one of the standard constructions to a proper Dupin submanifold with g distinct curvature spheres produces a proper Dupin submanifold with $g+1$ distinct curvature spheres defined on an open subset of $M^{n-1} \times S^m$. Pinkall [4, p.438] found the following simple criterion for reducibility in this general situation.

Theorem 2.8: *A proper Dupin submanifold* $\mu : W^{d-1} \to \Lambda^{2d-1}$ *with* $g+1 \geq 2$ *distinct curvature spheres is reducible to a proper Dupin submanifold with* g *distinct curvature spheres if and only if* μ *has a curvature sphere* $[K]$ *of multiplicity* $m \geq 1$ *which lies in a* $(d+1-m)$*–dimensional linear subspace of* $\mathbb{P}^{d+2}$.

Proof: Let $n = d-m$ in order to agree with the notation used earlier in this section. Since μ has at least 2 distinct curvature spheres, we have $d-1-m \geq 1$, i.e., $n \geq 2$. For each of the three constructions, it was shown that if μ has one more curvature sphere than the original Dupin submanifold λ, then the new curvature sphere $[K]$ has multiplicity m and lies in an $(d+1-m)$–dimensional linear subspace E of $\mathbb{P}^{d+2}$.

Conversely, if there exists a curvature sphere $[K]$ of multiplicity m which lies in an $(n+1)$–dimensional space E, then the signature of $< , >$ on $(m+1)$–dimensional vector space $E^{\perp}$ must be $(m+1, 0)$, $(m, 1)$ or $(m, 0)$. Otherwise, $E \cap Q^{d+1}$ is either empty or consists of a single point or a line (see Corollary 5.3 of Chapter 1). However, the curvature sphere map $[K]$ is an immersion of the $(n-1)$–dimensional space of leaves M^{n-1} of the principal foliation corresponding to $[K]$, and its image cannot be contained in a single line.

If the signature on $E^{\perp}$ is $(m+1, 0)$, then there is a Lie transformation A which takes $E^{\perp}$ to the space in (2.26). For the Dupin submanifold $A\mu$, the centers of the curvature spheres in the family [AK] all lie in

(2.27) $\mathbb{R}^{n-1}$ = Span $\{e_3,...,e_{n+1}\}$ $\subset$ $\mathbb{R}^n$ = Span $\{e_3,...,e_{n+2}\}$.

The proper Dupin submanifold $A\mu$ is an envelope of this family of curvature spheres [AK], with each curvature sphere tangent to the envelope along a leaf of the principal foliation corresponding to [AK]. Since the family of curvature spheres [AK] is invariant under $SO(m+1)$, the subgroup of $SO(d)$ consisting of isometries which keep the axis $\mathbb{R}^{n-1}$ pointwise fixed, the envelope of these curvature spheres is also invariant under $SO(m+1)$. Thus, $A\mu$ is an open subset of a surface of revolution. The profile submanifold λ in $\mathbb{R}^n$ of this surface of revolution is locally obtained by taking those contact elements in $\mathbb{R}^n$ which are in the image of $A\mu$. Each curvature surface of [AK] is the orbit of a contact element in the image of λ under the action of the group $SO(m+1)$. Since the multiplicity m of [AK] is accounted for by the action of this group, the profile submanifold has one less curvature sphere than μ at each point.

Similarly, if the signature of $E^{\perp}$ is $(m, 1)$, then $E^{\perp}$ can be mapped by a Lie transformation A to the space (2.14) in the tube construction. Then each curvature sphere in the family [AK] has radius $-\varepsilon$ and center in $\mathbb{R}^n$ (2.27). Since the map [AK] factors through an immersion of the space of leaves M^{n-1} of the principal foliation, the locus of centers of these spheres factors through an immersion f of M^{n-1} into $\mathbb{R}^n$. The proper Dupin submanifold $A\mu$ is an envelope of this family of curvature spheres, and it is obtained from the Legendre submanifold λ induced from the hypersurface f in $\mathbb{R}^n$ via the tube construction. Since the multiplicity of [AK] is accounted for by the tube construction, λ has one less curvature sphere than μ.

Finally, if $E^{\perp}$ has signature $(m, 0)$, then it can be taken by a Lie transformation A to the space (2.20). The family [AK] of curvature spheres consists of hyperplanes orthogonal to $\mathbb{R}^n$ (2.27). The proper Dupin submanifold $A\mu$ is an envelope of this family of hyperplanes, with each hyperplane tangent to the envelope along a leaf of the principal foliation. This family of hyperplanes is invariant under the action of the group H of translations of $\mathbb{R}^d$ in directions orthogonal to $\mathbb{R}^n$, and so is the envelope. Each leaf of the principal foliation is the orbit of a single contact element in $\mathbb{R}^n$

under the action of H. These contact elements in $\mathbb{R}^n$ determine the original proper Dupin submanifold λ from which $A\mu$ is obtained by the cylinder construction. Again, it is clear that λ has one less curvature sphere than μ at each point, since the multiplicity m of $[AK]$ equals the codimension of $\mathbb{R}^n$ in $\mathbb{R}^d$. $\qquad\square$

Pinkall also formulated his local criterion for reducibility to handle the case where the number of distinct curvature spheres of μ is the same as the number of distinct curvature spheres of λ, as well as the case above. For this theorem, we do not take into account the multiplicity of the curvature sphere $[K]$. The proof is essentially the same as that of the preceding theorem, and we omit it here. The result also holds for a proper Dupin submanifold with one curvature sphere at each point, so we include that case also.

Theorem 2.9: *A proper Dupin submanifold* $\mu : W^{d-1} \to \Lambda^{2d-1}$ *is reducible if and only if there is a curvature sphere* $[K]$ *of* μ *which lies in a d–dimensional linear subspace of* $\mathbb{P}^{d+2}$.

Remark 2.10: When Pinkall introduced his constructions, he also listed the following construction. Begin with the Dupin submanifold λ induced from an embedded Dupin hypersurface $M^{n-1} \subset S^n \subset \mathbb{R}^{n+1}$. The new Dupin submanifold μ is the Legendre submanifold induced from the cone C^n over M^{n-1} in $\mathbb{R}^{n+1}$ with vertex at the origin. The new family of curvature spheres consists of the hyperplanes tangent to the cone along its rulings. Theorem 2.9 shows that this construction is locally Lie equivalent to the tube construction as follows. The tube construction is characterized by the fact that one curvature sphere map $[K]$ lies in a d–dimensional linear subspace E of $\mathbb{P}^{d+2}$, whose orthogonal complement $E^{\perp}$ has signature $(1,1)$. Geometrically, this means that after a suitable Lie transformation, all the spheres in the family $[K]$ have the same radius and their centers lie in a subspace $\mathbb{R}^{d-1} \subset \mathbb{R}^d$. For the cone construction, the new family $[K]$ of curvature spheres consists of hyperplanes through the origin. Since the hyperplanes also all pass through the improper point, they correspond to points in the linear subspace E, whose orthogonal complement is as follows:

$$E^{\perp} = \text{Span } \{e_1 + e_2, e_1 - e_2\} \ .$$

Since $E^{\perp}$ is spanned by e_1 and e_2, it has signature $(1, 1)$. Thus, the cone construction is Lie equivalent to the tube construction. Finally, there is one more geometric interpretation of the tube construction. Note that a family $[K]$ of curvature spheres which lies in a linear subspace whose orthogonal complement has signature $(1, 1)$ can also be considered to consist of spheres in S^d of constant radius in the spherical metric whose centers lie in a hyperplane. The corresponding proper Dupin submanifold can thus be considered to be a tube in the spherical metric over a lower dimensional submanifold in S^d.

The considerations above are all of a local nature. We now want to consider the global question of when a compact proper Dupin hypersurface embedded in $\mathbb{R}^d$ is reducible to a compact Dupin hypersurface in a subspace $\mathbb{R}^n$ of $\mathbb{R}^d$. We say that a Dupin submanifold μ, obtained from a Dupin submanifold λ by one of the standard constructions, is *reducible to* λ . More generally, a Dupin submanifold η which is Lie equivalent to such a μ is reducible to λ. In terms of immersed hypersurfaces of Euclidean space, we say that a Dupin hypersurface $F : W^{d-1} \to \mathbb{R}^d$ is reducible to a Dupin hypersurface $f : M^{n-1} \to \mathbb{R}^n$, if the Legendre submanifold μ induced by F is reducible to the Legendre submanifold λ induced by f.

Thorbergsson [1] showed that a compact proper Dupin hypersurface immersed in $\mathbb{R}^n$ is taut, and therefore, it must be embedded (see Carter–West [1] or Cecil–Ryan [7, p.121]).

Theorem 2.11: *A compact, connected Dupin hypersurface W embedded in $\mathbb{R}^d$ with $g \geq 2$ distinct curvature spheres at each point is reducible to a compact proper Dupin hypersurface M embedded in $\mathbb{R}^n$ if and only if $g = 2$.*

Proof: If $g = 2$, then W must be a cyclide of Dupin (see Section 4.3 or Cecil–Ryan [2]), and therefore, it is reducible to a round sphere. Conversely, suppose that W is reducible to M. Since W is compact and M is immersed, we have from Propositions 2.1, 2.5 and 2.7 that W is diffeomorphic to $M \times S^m$, for some $m \geq 1$. Furthermore, if $g > 2$, then W must be obtained from M by the surface of revolution construction, since the other constructions always produce some points where the number of distinct curvature spheres is two. Proposition 2.7 also shows that the number of distinct curvature spheres on M must be g or $g-1$. We also have the following relationship between the sum β of the $\mathbb{Z}_2$–Betti numbers of W and M,

$$(2.28) \qquad \beta(W) = \beta(M \times S^m) = 2\,\beta(M)\ .$$

On the other hand, Thorbergsson [1] showed that for a compact proper Dupin hypersurface embedded in Euclidean space, β is equal to twice the number of distinct curvature spheres. This fact, together with (2.28), implies that it is impossible for W and M to have the same number of distinct curvature spheres, and thus M has $g-1$ distinct curvature spheres. Hence, we have

$$(2.29) \qquad \beta(W) = 2g\ , \qquad \beta(M) = 2(g-1) = 2g - 2\ .$$

Combining (2.28) and (2.29), we get

$$2g = 2(2g - 2) = 4g - 4\ ,$$

and thus $g = 2$, as desired.   □

A hypersurface in S^d is conformally equivalent to its image in $\mathbb{R}^d$ under stereographic projection. Furthermore, the proper Dupin condition is preserved under stereographic projection. Thus, as a corollary of Theorem 2.11, we conclude that a compact isoparametric hypersurface in S^d is reducible to a compact proper Dupin hypersurface in $S^n \subset S^d$ if and only if $g = 2$. Of course, compactness is not really a restriction for an isoparametric hypersurface, since Münzner [1] has shown that any connected piece of an isoparametric hypersurface lies in a unique compact, connected isoparametric hypersurface. There is one other geometric consequence about isoparametric hypersurfaces which is implied by the theorem. Münzner showed that an isoparametric hypersurface $M^{n-1} \subset S^n \subset \mathbb{R}^{n+1}$ is a tube of constant radius in S^n over each of its two focal submanifolds. If $g = 2$, then each of these focal submanifolds is a totally geodesic sphere of dimension $n-1-m$, where m is the multiplicity of the corresponding curvature sphere. On the other hand, if M^{n-1} has $g \geq 3$ distinct curvature spheres, then each focal submanifold must be substantial, i.e., it does not lie in any hyperplane in $\mathbb{R}^{n+1}$. Otherwise, there would be points on the tube at which there are only two distinct curvature spheres, as in Proposition 2.3.

Finally, Theorem 2.11 has the following corollary concerning the location of curvature spheres in $\mathbb{R}^n$.

Corollary 2.12: *Let $M^{n-1} \subset \mathbb{R}^n$ be a compact, connected proper Dupin hypersurface having $g \geq 2$ distinct curvature spheres at each point. Let $\mathbb{R}^{n-1}$ be any hyperplane in $\mathbb{R}^n$ which is disjoint from M^{n-1}. Then, there must exist a curvature sphere at some point of M^{n-1} which is orthogonal to $\mathbb{R}^{n-1}$.*

Proof: Consider $\mathbb{R}^n$ embedded as a subspace of a Euclidean space $\mathbb{R}^d$, $d > n$. Since M^{n-1} is disjoint from $\mathbb{R}^{n-1}$, the hypersurface W^{d-1} obtained by revolving M^{n-1} about $\mathbb{R}^{n-1}$ is embedded in $\mathbb{R}^d$. If no curvature sphere of M^{n-1} intersects $\mathbb{R}^{n-1}$ orthogonally, then by Proposition 2.7, W^{d-1} is a compact reducible proper Dupin hypersurface with more than two distinct curvature spheres, contradicting Theorem 2.11.   □

Geometrically, this corollary means that either the focal set of M^{n-1} in $\mathbb{R}^n$ must intersect $\mathbb{R}^{n-1}$, or there must be a point $x \in M^{n-1}$ where the tangent hyperplane to M^{n-1} is a curvature sphere orthogonal to $\mathbb{R}^{n-1}$. In that case, the corresponding focal point on the normal line to M^{n-1} at x intersects $\mathbb{R}^{n-1}$ at a point on the hyperplane at infinity in the projective space $\mathbb{P}^n$ determined by $\mathbb{R}^n$. In this sense, we can say that the focal set of M^{n-1} must intersect every hyperplane which is disjoint from M^{n-1}.

4.3 Cyclides of Dupin

A proper Dupin submanifold with two distinct curvature spheres of respective multiplicities p and q is called a cyclide of Dupin of *characteristic* (p, q). These are the simplest Dupin submanifolds after the spheres, and they were first studied in $\mathbb{R}^3$ by Dupin [1] in 1822. An example of a cyclide of Dupin is a torus of revolution. The long history of the classical cyclides in the nineteenth century is given in Lilienthal [1]. (See also Banchoff [1], Blaschke [1, p.238], Cecil [1], and Cecil–Ryan [7, pp.151–166] for more on the classical cyclides.) For Dupin cyclides in $\mathbb{R}^3$, it was known in the last century that every connected cyclide is Moebius equivalent to an open subset of a surface of revolution obtained by revolving a profile circle $S^1 \subset \mathbb{R}^2$ about an axis $\mathbb{R}^1 \subset \mathbb{R}^2 \subset \mathbb{R}^3$. The circle is allowed to intersect the axis, thereby introducing Euclidean singularities. However, the corresponding Legendre map into the space of contact elements on $\mathbb{R}^3$ is an immersion.

Higher dimensional cyclides appeared in the study of isoparametric hypersurfaces in spheres. Cartan knew that an isoparametric hypersurface in a sphere with two distinct curvature spheres at each point must be a standard product of spheres,

$$ S^p(r) \times S^q(s) \subset S^n(1) \subset \mathbb{R}^{p+1} \times \mathbb{R}^{q+1} = \mathbb{R}^{p+q+2}, \quad r^2 + s^2 = 1 . $$

Cecil and Ryan [2] showed that a compact proper Dupin hypersurface M^{n-1} embedded in S^n with two distinct curvature spheres must be Moebius equivalent to a standard product of spheres. The proof, however, uses the compactness of M^{n-1} in an essential way. Later, Pinkall [4] used Lie sphere geometry to obtain a local Lie geometric classification of the higher dimensional cyclides which is analogous to the classical result. In this section, we will prove Pinkall's theorem and then derive a local Moebius geometric classification from it. Pinkall's result is the following.

Theorem 3.1: (a) *Every connected cyclide of Dupin is contained in a unique compact, connected cyclide.*
(b) *Any two cyclides of Dupin of the same characteristic are locally Lie equivalent.*

Before proving the theorem, we consider some models for compact cyclides. The results of the previous section show that one can obtain a cyclide of characteristic (p, q) by applying any of the standard constructions to a p–sphere $S^p \subset \mathbb{R}^{p+1} \subset \mathbb{R}^n$, where $n = p+q+1$. Another simple model of a cyclide is obtained by considering the Legendre submanifold induced by a totally geodesic $S^q \subset S^n$, as a submanifold of codimension $p+1$. Such a sphere is one of the two focal submanifolds of the family of isoparametric hypersurfaces obtained by taking tubes over S^q in S^n. The other focal submanifold is a totally geodesic S^p in S^n. We now explicitly parametrize this Legendre submanifold by k_1 and k_2 satisfying the conditions (1) – (3) of Theorem 2.3 of Chapter 3. Let $e_1,...,e_{n+3}$ be the standard orthonormal basis for $\mathbb{R}_2^{n+3}$, and let

(3.1) $\Omega = \text{Span } \{e_1,...,e_{q+2}\}$, $\Omega^\perp = \text{Span } \{e_{q+3},...,e_{n+3}\}$.

These spaces have signatures $(q+1, 1)$ and $(p+1, 1)$, respectively. The intersection $\Omega \cap Q^{n+1}$ is given in homogeneous coordinates by

$$x_1^2 = x_2^2 + ... + x_{q+2}^2 , \qquad x_{q+3} = ... = x_{n+3} = 0 .$$

This set is diffeomorphic to the unit sphere S^q in $\mathbb{R}^{q+1} = \text{Span } \{e_2,...,e_{q+2}\}$ by the diffeomorphism $\varphi : S^q \to \Omega \cap Q^{n+1}$, $\varphi(v) = [e_1 + v]$. Similarly, $\Omega^\perp \cap Q^{n+1}$ is diffeomorphic to the unit sphere S^p in $\mathbb{R}^{p+1} = \text{Span } \{e_{q+3},...,e_{n+2}\}$ by the diffeomorphism $\psi : S^p \to \Omega^\perp \cap Q^{n+1}$, $\psi(u) = [u + e_{n+3}]$. The Legendre submanifold $\lambda : S^p \times S^q \to \Lambda^{2n-1}$ is defined by

(3.2) $\lambda(u, v) = [k_1, k_2]$, with $[k_1(u, v)] = [\varphi(v)]$, $[k_2(u, v)] = [\psi(u)]$.

It is easy to check that the conditions (1) – (3) are satisfied by $\{k_1, k_2\}$. To find the curvature spheres of λ, we decompose the tangent space to $S^p \times S^q$ at (u, v) as $T_u S^p \times T_v S^q$. Then, $dk_1(X, 0) = 0$ for all $X \in T_u S^p$, and $dk_2(0, Y) = 0$ for all Y in $T_v S^q$. Thus, $[k_1]$ and $[k_2]$ are curvature spheres of λ with respective multiplicities p and q. Furthermore, the image of $[k_1]$ lies in the set $\Omega \cap Q^{n+1}$, while the image of $[k_2]$ is contained in $\Omega^\perp \cap Q^{n+1}$. The essence of Pinkall's proof is to show that this type of relationship always holds between the two curvature spheres of a cyclide.

Proof of Theorem 3.1: Suppose that $\lambda : M^{n-1} \to \Lambda^{2n-1}$ is a connected cyclide of characteristic (p, q) with $p+q = n-1$. We may take $\lambda = [k_1, k_2]$, where $[k_1]$ and $[k_2]$ are the curvature spheres with respective multiplicities p and q. Each curvature sphere map factors through an immersion of the space of leaves of its principal foliation. Thus, locally on M^{n-1}, we can take a principal coordinate system (u, v) defined on an open set $W = U \times V \subset \mathbb{R}^p \times \mathbb{R}^q$ such that:

(i) $[k_1]$ depends only on v, and $[k_2]$ depends only on u, for $(u, v) \in W$.

(ii) $[k_1(W)]$ and $[k_2(W)]$ are submanifolds of Q^{n+1} of dimensions q and p, respectively.

Note that, in general, such a principal coordinate system cannot be found in the case of $g > 2$ curvature spheres. (See Cecil–Ryan [7, p.182].)

Now, let (u, v) and $(\bar{u}, \bar{v})$ be any two points in W. From (i), we have

$$(3.3) \quad < k_1(u, v), k_2(\bar{u}, \bar{v}) > \ = \ < k_1(v), k_2(\bar{u}) > \ = \ < k_1(\bar{u}, v), k_2(\bar{u}, v) > \ = \ 0 .$$

Let E be the smallest linear subspace of $\mathbb{P}^{n+2}$ containing the q–dimensional submanifold $[k_1(W)]$. By (3.3), we have

$$(3.4) \qquad [k_1(W)] \subset E \cap Q^{n+1}, \qquad [k_2(W)] \subset E^\perp \cap Q^{n+1}.$$

The dimensions of E and $E^\perp$ as subspaces of $\mathbb{P}^{n+2}$ satisfy

$$(3.5) \qquad \dim E + \dim E^\perp = n + 1 = p + q + 2 .$$

We claim that $\dim E = q + 1$ and $\dim E^\perp = p + 1$. To see this, suppose first that $\dim E > q + 1$. Then, $\dim E^\perp \leq p$, and $E^\perp \cap Q^{n+1}$ cannot contain the p–dimensional submanifold $k_2(W)$. Similarly, assuming that $\dim E^\perp > p + 1$ leads to a contradiction. Thus, we have $\dim E \leq q + 1$ and $\dim E^\perp \leq p + 1$. This and (3.5) imply that $\dim E = q + 1$ and $\dim E^\perp = p + 1$. Furthermore, from the fact that $E \cap Q^{n+1}$ and $E^\perp \cap Q^{n+1}$ contain submanifolds of dimensions q and p, respectively, it is easy to deduce that $< , >$ has signature $(q + 1, 1)$ on

E and $(p + 1, 1)$ on $E^{\perp}$. Then, since $E \cap Q^{n+1}$ and $E^{\perp} \cap Q^{n+1}$ are diffeomorphic to S^q and S^p, respectively, the inclusions in (3.4) are open subsets. If A is a Lie transformation which takes E to the space Ω in (3.1), and thus takes $E^{\perp}$ to $\Omega^{\perp}$, then $A\lambda(W)$ is an open subset of the standard model in (3.2). Both assertions of Theorem 3.1 are now clear.   □

We now turn to the Moebius geometric classification of the cyclides. For the classical cyclides in $\mathbb{R}^3$, this was known in the last century. At Oberwolfach in 1981, K. Voss [1] announced the classification in Theorem 3.2 below for the higher dimensional cyclides, but he did not publish a proof. The theorem follows quite directly from Theorem 3.1 and the results of the previous section on surfaces of revolution. The theorem is phrased in terms of embedded hypersurfaces in $\mathbb{R}^n$. Thus, we are excluding the standard model (3.2), where the Euclidean projection is not an immersion. Of course, the Euclidean projection of a parallel submanifold to the standard model is an embedding. This proof was also given in Cecil [4].

Theorem 3.2: (a) *Every connected cyclide of Dupin M^{n-1} of characteristic (p, q) embedded in $\mathbb{R}^n$ is Moebius equivalent to an open subset of a surface of revolution obtained by revolving a q-sphere $S^q \subset \mathbb{R}^{q+1} \subset \mathbb{R}^n$ about an axis of revolution $\mathbb{R}^q \subset \mathbb{R}^{q+1}$ or a p-sphere $S^p \subset \mathbb{R}^{p+1} \subset \mathbb{R}^n$ about an axis $\mathbb{R}^p \subset \mathbb{R}^{p+1}$.*
(b) *Two such surfaces are Moebius equivalent if and only if they have the same value of $\rho = |r| / a$, where r is the signed radius of the profile sphere S^q and $a > 0$ is the distance from the center of S^q to the axis of revolution.*

Proof: We always work with the Legendre submanifold induced by the embedding of M^{n-1} into $\mathbb{R}^n$. By Theorem 3.1, every connected cyclide is contained in a unique compact, connected cyclide. Thus, it suffices to classify compact, connected cyclides up to Moebius equivalence. Consider a compact, connected cyclide $\lambda : S^p \times S^q \to \Lambda^{2n-1}$ of characteristic (p, q). By Theorem 3.1, there is a linear subspace E of $\mathbb{P}^{n+2}$ with signature $(q+1, 1)$ such that the two curvature sphere maps, $[k_1] : S^q \to E \cap Q^{n+1}$ and $[k_2] : S^p \to E^{\perp} \cap Q^{n+1}$, are diffeomorphisms.

Moebius transformations are precisely those Lie transformations satisfying $A[e_{n+3}] = [e_{n+3}]$. Thus, we decompose e_{n+3} as

(3.6) $e_{n+3} = \alpha + \beta$, $\alpha \in E$, $\beta \in E^{\perp}$.

Note that since $< \alpha, \beta > = 0$, we have

$$-1 = < e_{n+3} , e_{n+3} > = < \alpha, \alpha > + < \beta, \beta > .$$

Hence, at least one of the two vectors α, β is timelike. First, suppose that β is timelike. Let Z be the orthogonal complement of β in $E^{\perp}$. Then Z is a $(p+1)$–dimensional vector space with signature $(p+1, 0)$. Since $Z \subset e_{n+3}^{\perp}$, there is a Moebius transformation A such that

$$A(Z) = S = \text{Span } \{e_{q+3},...,e_{n+2}\} .$$

The curvature sphere map $[Ak_1]$ of $A\lambda$ is a q–dimensional submanifold in the space $S^{\perp} \cap Q^{n+1}$. By (3.4) of Chapter 1, this means that these spheres all have their centers in the space $\mathbb{R}^q = \text{Span } \{e_3,...,e_{q+2}\}$. Note that

$$\mathbb{R}^q \subset \mathbb{R}^{q+1} = \text{Span } \{e_3,...,e_{q+3}\} \subset \mathbb{R}^n = \text{Span } \{e_3,...,e_{n+2}\} .$$

As we see from the proof of Theorem 2.8, this means that the Dupin submanifold $A\lambda$ is a surface of revolution in $\mathbb{R}^n$ obtained by revolving a q–dimensional profile submanifold in $\mathbb{R}^{q+1}$ about the axis $\mathbb{R}^q$. Moreover, since $A\lambda$ has two distinct curvature spheres, the profile submanifold has only one curvature sphere. Thus, it is an umbilic submanifold of $\mathbb{R}^{q+1}$.

Four cases are naturally distinguished by the nature of the vector α in (3.6). Geometrically, these correspond to different singularity sets of the Euclidean projection of $A\lambda$. Such singularities correspond exactly with the singularities of the Euclidean projection of λ, since the Moebius transformation A preserves the rank of the Euclidean projection. Since we have assumed that β is timelike, we know that for all $u \in S^p$,

$$< k_2(u), e_{n+3} > = < k_2(u), \alpha + \beta > = < k_2(u), \beta > \neq 0 ,$$

because the orthogonal complement of β in $E^\perp$ is spacelike. Thus, the curvature sphere $[Ak_2]$ is never a point sphere. However, it is possible for $[Ak_1]$ to be a point sphere.

Case 1: $\alpha = 0$. In this case, the curvature sphere $[Ak_1]$ is a point sphere for every point in $S^p \times S^q$. The image of the Euclidean projection of $A\lambda$ is precisely the axis $\mathbb{R}^q$. The cyclide $A\lambda$ is the Legendre submanifold induced from $\mathbb{R}^q$ as a submanifold of codimension $p+1$ in $\mathbb{R}^n$. This is, in fact, the standard model. However, since the Euclidean projection is not an immersion, this case does not lead to any of the embedded hypersurfaces classified in (a).

In the remaining cases, we can always arrange that the umbilic profile submanifold is a q–sphere and not a q–plane. This can be accomplished by first inverting $\mathbb{R}^{q+1}$ in a sphere centered at a point on the axis $\mathbb{R}^q$ which is not on the profile submanifold, if necessary. Such an inversion preserves the axis of revolution $\mathbb{R}^q$. After a Euclidean translation, we may assume that the center of the profile sphere is a point $(0, a)$ on the x_{q+3}–axis ℓ in $\mathbb{R}^{q+1}$, as in Figure 3.1. The center of the profile sphere cannot lie on the axis of revolution $\mathbb{R}^q$, for then the surface of revolution would be an $(n-1)$–sphere and not a cyclide. Thus, we may take $a > 0$.

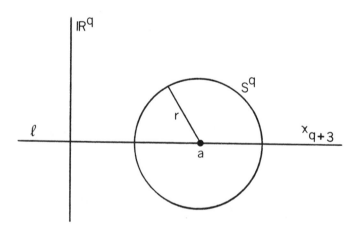

Figure 3.1 – Profile sphere S^q for the surface of revolution

The map $[Ak_1]$ is the curvature sphere of $A\lambda$ which results from the surface of revolution construction. The other curvature sphere of $A\lambda$ corresponds exactly to the curvature sphere of the profile sphere, i.e., to the profile sphere itself. This means that the signed radius r of the profile sphere is equal to the signed radius of the curvature sphere $[Ak_2]$. Since $[Ak_2]$ is never a point sphere, $r \neq 0$. From now on, we will identify the profile sphere with the second factor S^q in the domain of λ.

Case 2: α *is timelike.* In this case, for all $v \in S^q$,

$$< k_1(v), e_{n+3} > \ = \ < k_1(v), \alpha > \ \neq 0 \ ,$$

since the orthogonal complement of α in E is spacelike. Thus, the Euclidean projection of $A\lambda$ is an immersion at all points. This corresponds to the case $|r| < a$, when the profile sphere is disjoint from the axis of revolution. Classically, these were known as the *ring cyclides* (see Cecil–Ryan [7]). Note that by interchanging the roles of α and β, we can find a Moebius transformation which takes λ to the Legendre submanifold obtained by revolving a p–sphere around an axis $\mathbb{R}^p \subset \mathbb{R}^{p+1} \subset \mathbb{R}^n$.

Case 3: α *is lightlike, but not zero.* Then, there is exactly one $v \in S^q$ such that

$$(3.7) \qquad\qquad < k_1(v), e_{n+3} > \ = \ < k_1(v), \alpha > \ = 0 \ .$$

This corresponds to the case $|r| = a$, where the profile sphere intersects the axis in one point. Thus, $S^p \times \{v\}$ is the set of points in $S^p \times S^q$ where the Euclidean projection is singular. Classically, these were known as *limit spindle cyclides.*

Case 4: α *is spacelike.* Then, the condition (3.7) holds for points v in a $(q-1)$–sphere $S^{q-1} \subset S^q$. For points in $S^p \times S^{q-1}$, the point sphere map is a curvature sphere, and thus the Euclidean projection is singular. Geometrically, this is the case $|r| > a$, where the profile sphere intersects the axis $\mathbb{R}^q$ in a $(q-1)$–sphere. Classically, these were known as *spindle cyclides.*

Of course, there are also four cases to handle under the assumption that α, instead of β, is timelike. Then, the axis will be a subspace $\mathbb{R}^p \subset \mathbb{R}^{p+1}$, and the profile submanifold will be a p–sphere. The roles of p and q in determining the dimension of the singularity set of the Euclidean projection will be reversed. So if $p \neq q$, then only a ring cyclide can be represented as a surface of revolution of both a q–sphere and a p–sphere.

To prove (b), we may assume that the profile sphere S^q of the surface of revolution has center $(0, a)$, $a > 0$, on the x_{q+3}–axis ℓ. Moebius classification clearly does not depend on the sign of the radius of S^q, since the two surfaces of revolution obtained by revolving spheres with the same center and opposite radii differ only by the change of orientation transformation Γ (see Remark 1.4 of Chapter 2). We now show that the ratio $\rho = |r| \, / \, a$ is invariant under the subgroup of Moebius transformations of the profile space $\mathbb{R}^{q+1}$ which take one such surface of revolution to another. First, note that symmetry implies that a transformation T in this subgroup must take the axis of revolution $\mathbb{R}^q$ to itself and the axis of symmetry ℓ to itself. Since $\mathbb{R}^q$ and ℓ intersect only at 0 and ∞, the transformation T maps the set $\{0, \infty\}$ to itself. If T maps ∞ to 0, then the composition ΦT, where Φ is an inversion in a sphere centered at 0, is a member of the subgroup which takes ∞ to ∞ and 0 to 0. By Theorem 5.1 of Chapter 2, such a Moebius transformation must be a similarity transformation, and so it is the composition of a central dilatation D with a linear isometry Ψ. Therefore, $T = \Phi D \Psi$, and each of the transformations on the right of this equation preserves the ratio ρ. The invariant ρ is the only one needed for Moebius classification, since any two profile spheres with the same value of ρ can be mapped to one another by a central dilatation. $\square$

Remark 3.3: We can obtain a family consisting of one representative from each Moebius equivalence class by fixing $a = 1$ and letting r vary, $0 < r < \infty$. This is just a family of parallel hypersurfaces of revolution. Taking a negative signed radius s for the profile sphere yields a parallel hypersurface which differs only in orientation from the surface corresponding to $r = -s$. Finally, taking $r = 0$ also gives a parallel submanifold in the family, but the Euclidean projection degenerates to a sphere S^p. This is the case $\beta = 0$, $\alpha = e_{n+3}$, where the point sphere map equals the curvature sphere $[k_2]$ at every point.

4.4 Principal Lie Frames

The goal of this section is to construct Lie frames which are well suited for the local study of Dupin submanifolds. These principal frames are fundamental in the local classification of Dupin hypersurfaces in $\mathbb{R}^4$ given in Section 4.6. They also provide a logical starting point for further local study of Dupin submanifolds. Moreover, a principal frame can be constructed locally for any Legendre submanifold. It is only in the last part of this section that we invoke the Dupin condition.

Let $\lambda : M^{n-1} \to \Lambda^{2n-1}$ be an arbitrary Legendre submanifold. Let $\{Y_a\}$ be a smooth Lie frame on an open subset U of M^{n-1} such that for each $x \in M^{n-1}$, we have $\lambda(x) = [Y_1(x), Y_{n+3}(x)]$. We will again use the notation for the method of moving Lie frames introduced in Section 3.1. In particular, we agree on the following range of indices,

(4.1) $$ 1 \leq a, b, c \leq n+3 , \qquad 3 \leq i, j, k \leq n+1 . $$

All summations are over the repeated index or indices. Recall that the Maurer–Cartan forms ω_a^b are defined by the equation

(4.2) $$ dY_a = \sum \omega_a^b Y_b . $$

We will pull back these structure forms to M^{n-1} and omit the symbols of such pull–backs for simplicity. Recall that the following matrix of forms is skew–symmetric,

(4.3)
$$
\begin{bmatrix}
\omega_1^2 & \omega_1^1 & \omega_1^i & \omega_1^{n+3} & \omega_1^{n+2} \\
\omega_2^2 & \omega_2^1 & \omega_2^i & \omega_2^{n+3} & \omega_2^{n+2} \\
\omega_j^2 & \omega_j^1 & \omega_j^i & \omega_j^{n+3} & \omega_j^{n+2} \\
\omega_{n+2}^2 & \omega_{n+2}^1 & \omega_{n+2}^i & \omega_{n+2}^{n+3} & \omega_{n+2}^{n+2} \\
\omega_{n+3}^2 & \omega_{n+3}^1 & \omega_{n+3}^i & \omega_{n+3}^{n+3} & \omega_{n+3}^{n+2}
\end{bmatrix}
$$

and that the forms satisfy the Maurer–Cartan equations,

$$(4.4) \qquad\qquad d\omega_a^b = \sum \omega_a^c \wedge \omega_c^b \ .$$

By Theorem 4.4 of Chapter 3, there are at most $n-1$ distinct curvature spheres along each line $\lambda(x)$. Thus, we can choose the Lie frame locally so that neither Y_1 nor Y_{n+3} is a curvature sphere at any point. Now, $\omega_1^2 = 0$ by the skew–symmetry of (4.3), and $\omega_1^{n+2} = 0$ by the Legendre condition (3) for λ. Thus, for any $X \in T_x M^{n-1}$ at any point $x \in U$, we have

$$(4.5) \quad dY_1(X) = \omega_1^1(X)Y_1 + \sum \omega_1^i(X)Y_i + \omega_1^{n+3}(X) \equiv \sum \omega_1^i(X)Y_i , \quad \text{mod } \{Y_1, Y_{n+3}\} \ .$$

The assumption that Y_1 is not a curvature sphere means that there does not exist any non–zero tangent vector X at any point $x \in U$ such that $dY_1(X)$ is congruent to zero, mod $\{Y_1(x), Y_{n+3}(x)\}$. By (4.5), this assumption is equivalent to the condition that the forms $\omega_1^3,...,\omega_1^{n+1}$ be linearly independent, i.e., they satisfy the regularity condition,

$$(4.6) \qquad\qquad \omega_1^3 \wedge ... \wedge \omega_1^{n+1} \neq 0 \ ,$$

on U. Similarly, the assumption that Y_{n+3} is not a curvature sphere is equivalent to the condition

$$(4.7) \qquad\qquad \omega_{n+3}^3 \wedge ... \wedge \omega_{n+3}^{n+1} \neq 0 \ .$$

We want to construct a Lie frame which is well–suited for the study of the curvature spheres of λ. The Legendre condition (3) for λ is $\omega_1^{n+2} = 0$. Exterior differentiation of this equation, using (4.3) − (4.4), yields

$$(4.8) \qquad\qquad \sum \omega_1^i \wedge \omega_{n+3}^i = 0 \ .$$

Hence, by Cartan's Lemma and the linear independence condition (4.6), we get that for each i,

(4.9) $$\omega_{n+3}^{i} = \sum h_{ij}\, \omega_1^{j}, \quad \text{with } h_{ij} = h_{ji}.$$

The quadratic differential form

(4.10) $$II\,(Y_1) = \sum h_{ij}\, \omega_1^{i}\, \omega_1^{j},$$

defined up to a non–zero factor and dependent on the choice of Y_1, is called the *second fundamental form* of λ determined by Y_1.

To justify this name, we now consider the special case where Y_1 and Y_{n+3} are as follows:

(4.11) $$Y_1 = (\,1 + f \cdot f,\, 1 - f \cdot f,\, 2f,\, 0\,)\,/\,2, \quad Y_{n+3} = (f \cdot \xi,\, -f \cdot \xi,\, \xi,\, 1\,),$$

where f is the Euclidean projection of λ, and ξ is the Euclidean field of unit normals. The condition (4.6) is equivalent to assuming that f is an immersion on U. Since f is an immersion, we can choose the Lie frame vectors $Y_3, \dots, Y_{n+1}$ to satisfy

(4.12) $$Y_i = dY_1(\,X_i\,) = (\,f \cdot df(\,X_i\,),\, -f \cdot df(\,X_i\,),\, df(\,X_i\,),\, 0\,), \quad 3 \le i \le n+1,$$

where $X_3, \dots, X_{n+1}$ are smooth vector fields on U. Then, we have

(4.13) $$\omega_1^{i}(\,X_j\,) = \langle\, dY_1(\,X_j\,),\, Y_i\, \rangle = \langle\, Y_j,\, Y_i\, \rangle = \delta_{ij}.$$

Using (4.11) and (4.12), we compute

(4.14) $$\omega_{n+3}^{i}(X_j) = \langle\, dY_{n+3}(X_j),\, Y_i\, \rangle = d\xi(X_j) \cdot df(X_i) = -\,df(AX_j) \cdot df(X_i) = -\,A_{ij}$$

where $[A_{ij}]$ is the Euclidean shape operator (second fundamental form) of f. Now by (4.9) and (4.13), we have

(4.15) $$\omega_{n+3}^{i}(\,X_j\,) = \sum h_{ik}\, \omega_1^{k}(\,X_j\,) = h_{ij}.$$

Thus, $h_{ij} = -\,A_{ij}$, whence the name "second fundamental form" for $[h_{ij}]$.

We now return to our general discussion, where λ is an arbitrary Legendre submanifold, and $\{Y_a\}$ is a Lie frame on U such that Y_1 and Y_{n+3} satisfy (4.6) and (4.7), respectively. Since $[h_{ij}]$ is symmetric, we can diagonalize it at any given point $x \in U$ by a change of frame of the form

$$Y_i^* = \Sigma C_i^j Y_j , \quad 3 \le i \le n+1 ,$$

where $[C_i^j]$ is an $(n-1) \times (n-1)$ orthogonal matrix. In the new frame, equation (4.9) has the following form at x,

(4.16) $$\omega_{n+3}^i = -\mu_i \omega_1^i , \quad 3 \le i \le n+1 .$$

These μ_i determine the curvature spheres of λ at x. Specifically, given any point $x \in U$, let $\{X_3,...,X_{n+1}\}$ be the dual basis to $\{\omega_1^3,...,\omega_1^{n+1}\}$ in the tangent space $T_x M^{n-1}$. Then using (4.16), we compute the differential of $\mu_i Y_1 + Y_{n+3}$ on X_i to be

$$d(\mu_i Y_1 + Y_{n+3})(X_i) = d\mu_i(X_i) Y_1 + (\mu_i dY_1 + dY_{n+3})(X_i)$$
$$\equiv \Sigma(\mu_i \omega_1^j(X_i) + \omega_{n+3}^j(X_i)) Y_j$$
$$= (\mu_i \omega_1^i(X_i) + \omega_{n+3}^i(X_i)) Y_i = (\mu_i - \mu_i) Y_i = 0 , \quad \text{mod } \{Y_1, Y_{n+3}\} .$$

Hence, the curvature spheres of λ at x are precisely

(4.17) $$K_i = \mu_i Y_1 + Y_{n+3} , \quad 3 \le i \le n+1 ,$$

and $X_3,...,X_{n+1}$ are the principal vectors at x. In the case where Y_1 and Y_{n+3} have the form (4.11), the μ_i are just the principal curvatures of the immersion f at the point x. From Corollary 4.5 of Chapter 3, we know that if a curvature sphere of the form (4.17) has constant multiplicity on U, then K_i and μ_i are both smooth on U, and the corresponding distribution of principal vectors is a foliation. As we noted after the proof of that corollary, there is an open, dense subset of M^{n-1} on which the multiplicities of the curvature spheres are locally constant (see Reckziegel [1]–[2] or Singley [1]). For the remainder of this section, we assume that the number g of distinct curvature spheres is constant on U, and thus, that each distinct curvature sphere has constant multiplicity on U. Assuming such, the principal vector fields $X_3,...,X_{n+1}$ can be chosen to be smooth on U. The frame vectors $Y_3,...,Y_{n+1}$ can then be chosen to be smooth on U via the formula

$$Y_i = dY_1(X_i), \quad 3 \le i \le n+1 .$$

As in (4.13), this means that $\{\omega_1^3,...,\omega_1^{n+1}\}$ is the dual basis to $\{X_3,...,X_{n+1}\}$. The equation (4.16) is then satisfied at every point of U. This frame is an example of what we will call a principal frame. In general, a Lie frame $\{Z_a\}$ on U is called a *principal Lie frame* if there exist smooth functions α_i and β_i on U, which are never simultaneously zero, such that the Maurer–Cartan forms $\{\theta_a^b\}$ for the frame satisfy the equations

$$\alpha_i \theta_1^i + \beta_i \theta_{n+3}^i = 0, \quad 3 \le i \le n+1 .$$

It may be worth noting that θ_1^i and θ_{n+3}^i cannot both vanish at a point x in U. To see this, take a Lie frame $\{W_a\}$ on U with $W_i = Z_i$, $3 \le i \le n+1$, such that $W_1 = \alpha Z_1 + \beta Z_{n+3}$ is not a curvature sphere at x. Then, the Maurer–Cartan form φ_1^i for this frame satisfies

$$\varphi_1^i = \; < dW_1, W_i > \; = \; < \alpha \, dZ_1 + \beta \, dZ_{n+3}, Z_i > \; = \alpha \, \theta_1^i + \beta \, \theta_{n+3}^i .$$

Since W_1 is not a curvature sphere, it follows that $\varphi_1^i \neq 0$, and thus, it is not possible for θ_1^i and θ_{n+3}^i to both equal zero.

There is a fair amount of flexibility in the choice of a principal Lie frame. We next show how the choice can be made more specific in order to have the Maurer–Cartan forms give more direct information about a given curvature sphere. We assume that $\{Y_a\}$ is a principal frame on U satisfying (4.6) – (4.7) and that the curvature spheres are given by (4.17). In particular, suppose $K = \mu Y_1 + Y_{n+3}$ is a curvature sphere of multiplicity m on U. Then, the function μ is smooth, and we can re-order the frame vectors $Y_3,...,Y_{n+1}$ so that

(4.18) $\mu = \mu_3 = \ldots = \mu_{m+2}$

on U. Since Y_1 and Y_{n+3} are not curvature spheres at any point of U, the function μ never takes the value 0 or ∞ on U. We now make a change of frame so that $\overset{*}{Y}_1 = K$ is a curvature sphere of multiplicity m. Specifically, let

$$\begin{aligned} \overset{*}{Y}_1 &= \mu Y_1 + Y_{n+3}, & \overset{*}{Y}_2 &= (1/\mu)\, Y_2, \\ \overset{*}{Y}_{n+2} &= Y_{n+2} - (1/\mu)\, Y_2, & \overset{*}{Y}_{n+3} &= Y_{n+3}, \\ \overset{*}{Y}_i &= Y_i, \quad 3 \le i \le n+1. \end{aligned}$$

(4.19)

Denote the Maurer–Cartan forms for this frame by θ_a^b. Note that

(4.20) $d\overset{*}{Y}_1 = d(\mu Y_1 + Y_{n+3}) = (d\mu)Y_1 + \mu\, dY_1 + dY_{n+3} = \sum \theta_1^a\, \overset{*}{Y}_a.$

Using (4.16), we see that the coefficient of $\overset{*}{Y}_i = Y_i$ in (4.20) is

(4.21) $\theta_1^i = \mu\, \omega_1^i + \omega_{n+3}^i = (\mu - \mu_i)\, \omega_1^i, \quad 3 \le i \le n+1.$

This and (4.18) show that

(4.22) $\theta_1^r = 0, \quad 3 \le r \le m+2.$

Equation (4.22) characterizes the condition that $\overset{*}{Y}_1$ is a curvature sphere of constant multiplicity m on U.

We now want to study the Dupin condition that $K = \overset{*}{Y}_1$ is constant along the leaves of its principal foliation. As noted in Corollary 4.5 of Chapter 3, this is automatic if the multiplicity m of K is greater than one. We denote this principal foliation by T_1 and choose smooth vector fields $X_3,\ldots,X_{m+2}$ on U which span T_1. The condition that $\overset{*}{Y}_1$ be constant along the leaves of its principal foliation is

(4.23) $d\overset{*}{Y}_1(X_r) \equiv 0, \quad \mathrm{mod}\ \overset{*}{Y}_1, \quad 3 \le r \le m+2.$

On the other hand, from (4.13) and (4.21), we have

(4.24) $dY_1^*(X_r) = \theta_1^1(X_r) Y_1 + \theta_1^{n+3}(X_r) Y_{n+3}$, $3 \le r \le m+2$.

Comparing (4.23) and (4.24), we see that

(4.25) $\theta_1^{n+3}(X_r) = 0$, $3 \le r \le m+2$.

We now show that we can make one more change of frame so that in the new frame the Maurer–Cartan form $\alpha_1^{n+3} = 0$. We can write θ_1^{n+3} in terms of the basis $\omega_1^3, ..., \omega_1^{n+1}$ as

(4.26) $\theta_1^{n+3} = \sum s_i \omega_1^i$,

for smooth functions s_i on U. From (4.25), we see that we actually have

(4.27) $\theta_1^{n+3} = \sum_{m+3}^{n+1} s_t \omega_1^t$.

Using (4.13), (4.21) and (4.27), we compute that for $m+3 \le t \le n+1$,

(4.28)
$$\begin{aligned}
dY_1^*(X_t) &= \theta_1^1(X_t) Y_1^* + \theta_1^t(X_t) Y_t + \theta_1^{n+3}(X_t) Y_{n+3} \\
&= \theta_1^1(X_t) Y_1^* + (\mu-\mu_t) Y_t + s_t Y_{n+3} \\
&= \theta_1^1(X_t) Y_1^* + (\mu-\mu_t)(Y_t + (s_t/(\mu-\mu_t))) Y_{n+3} .
\end{aligned}$$

We now make the change of Lie frame,

$Z_1 = Y_1^*$, $Z_2 = Y_2^*$, $Z_{n+3} = Y_{n+3}^*$, $Z_r = Y_r^* = Y_r$, $3 \le r \le m+2$,
(4.29) $Z_t = Y_t + (s_t/(\mu-\mu_t)) Y_{n+3}$, $m+3 \le t \le n+1$,
$Z_{n+2} = -\sum_t (s_t/(\mu-\mu_t)) Y_t + Y_{n+2} - (1/2)\sum_t (s_t/(\mu-\mu_t))^2 Y_{n+3}$.

Let α_a^b be the Maurer–Cartan forms for this new frame. We still have

(4.30) $\alpha_1^r = <dZ_1, Z_r> = <dY_1^*, Y_r^*> = \theta_1^r = 0$, $3 \le r \le m+2$.

Furthermore, since $Z_1 = Y_1^*$, the Dupin condition (4.23) still yields

$$\alpha_1^{n+3}(X_r) = 0, \quad 3 \le r \le m+2.$$

Finally, for $m+3 \le t \le n+1$, equations (4.28) and (4.29) yield

$$(4.31) \quad \alpha_1^{n+3}(X_t) = <dZ_1(X_t), Z_{n+2}> = <\theta_1^1(X_t)Z_1 + (\mu-\mu_t)Z_t, Z_{n+2}> = 0.$$

Thus, we have $\alpha_1^{n+3} = 0$. We summarize these results in the following theorem.

Theorem 4.1: *Let* $\lambda : M^{n-1} \to \Lambda^{2n-1}$ *be a Legendre submanifold. Suppose that K is a curvature sphere of multiplicity m on an open subset U of* M^{n-1} *which is constant along its curvature surfaces. Then locally on U, there exists a Lie frame* $\{Y_1,...,Y_{n+3}\}$ *with* $Y_1 = K$, *such that the Maurer-Cartan forms satisfy*

$$\omega_1^r = 0, \quad 3 \le r \le m+2, \quad \omega_1^{n+3} = 0.$$

4.5 Half–Invariant Differentiation

In this section, we discuss the method of "half–invariant differentiation" which is sometimes useful in simplifying the computation of derivatives in the projective context. This method is discussed in complete generality in Bol [1, Vol. III, pp.1–15]. Here we show how it can be used in the study of Legendre submanifolds. The method of half–invariant differentiation plays an important role in the local classification of Dupin hypersurfaces in $\mathbb{R}^4$ given in the next section, and it has been used by Niebergall [1] in his partial classification of Dupin hypersurfaces in $\mathbb{R}^5$.

Let $\lambda : M^{n-1} \to \Lambda^{2n-1}$ be a Legendre submanifold, and suppose that $\{Y_a\}$ is a Lie frame defined on an open subset U of M^{n-1}. One degree of freedom that is always present in the choice of a Lie frame is a *renormalization* of the form

$$\overset{*}{Y}_1 = \tau \, Y_1, \qquad\qquad \overset{*}{Y}_2 = (1/\tau) \, Y_2,$$

(5.1) $\qquad \overset{*}{Y}_{n+3} = \tau \, Y_{n+3}, \qquad\qquad \overset{*}{Y}_{n+2} = (1/\tau) \, Y_{n+2},$

$$\overset{*}{Y}_i = Y_i, \quad 3 \le i \le n+1,$$

where τ is a smooth non–vanishing function on U. This is equivalent to a "star–transformation", as defined by Pinkall [1], [3]. If Y_1 is not a curvature sphere at any point of U, then the forms $\{\omega_1^3,...,\omega_1^{n+1}\}$ comprise a basis for the cotangent space T^*U at each point of U. For a principal frame, it may be that some of the ω_1^i are zero. In that case, as we showed in the previous section, the corresponding form ω_{n+3}^i is not zero, and one can still choose a basis locally for T^*U so that each element of the basis is either an ω_1^i or an ω_{n+3}^i. We assume that we have such a basis, which we denote by $\{\theta_1,...,\theta_{n-1}\}$. Under a renormalization of the type (5.1), we have for $3 \le i \le n+1$,

$$\omega_1^{i*} = \; < d\overset{*}{Y}_1, \overset{*}{Y}_i > \; = \; < d(\tau Y_1), Y_i > \; = \; < (d\tau)Y_1 + \tau \, dY_1, Y_i >$$
$$= \tau < dY_1, Y_i > \; = \tau \, \omega_1^i.$$

In a similar way, $\omega_{n+3}^{i*} = \tau \, \omega_{n+3}^i$, for $3 \le i \le n+1$. Thus, all of the elements of the basis transform by the formula,

(5.2) $\qquad\qquad\qquad \overset{*}{\theta}_r = \tau \, \theta_r, \quad 1 \le r \le n-1.$

The form ω_1^1 transforms as follows under renormalization,

(5.3)
$$\omega_1^{1*} = \; < d\overset{*}{Y}_1, \overset{*}{Y}_2 > \; = \; < d(\tau Y_1), (1/\tau)Y_2 > \; = (1/\tau) < \tau \, dY_1 + (d\tau)Y_1, Y_2 >$$
$$= \; < dY_1, Y_2 > + \, (d\tau/\tau) < Y_1, Y_2 > \; = \omega_1^1 + d(\log \tau).$$

Hence, the form $\pi = -\omega_1^1$ transforms as a *Reeb form* for the renormalization (Bol [1]), i.e.,

(5.4) $\qquad\qquad\qquad \overset{*}{\pi} = \pi - d(\log \tau).$

Such a Reeb form is crucial in the half–invariant differentiation process.

A scalar function or differential form ω on U is said to be *half-invariant of weight m* if there is an integer m such that under renormalization, ω transforms by the formula

(5.5) $$\omega^* = \tau^m \omega .$$

Thus, the basis forms θ_r, $1 \le r \le n-1$, are all half–invariant of weight 1. For two half–invariant forms, ω_1 and ω_2, we set

$$(\omega_1 \wedge \omega_2)^* = \omega_1^* \wedge \omega_2^* .$$

If ω_1 has weight m_1 and ω_2 has weight m_2, then by (5.5) the product $\omega_1 \wedge \omega_2$ has weight $m_1 + m_2$. This is useful in determining the weight of certain functions which arise naturally. For example, if $3 \le i \le n-1$, then the form ω_i^1 has weight -1, since

$$\omega_i^{1*} = < dY_i^*, Y_2 > = < dY_i, (1/\tau)Y_2 > = (1/\tau) < dY_i, Y_2 > = (1/\tau)\, \omega_i^1 .$$

Now, suppose we write ω_i^1 in terms of the basis $\{\theta_r\}$ as

$$\omega_i^1 = a_1\theta_1 + ... + a_{n-1}\theta_{n-1} .$$

Then, each of the functions a_r has weight -2, since the weight of a_r plus the weight of θ_r must be -1.

The *half-invariant derivative* $\tilde{d}\omega$ of a function or form ω of weight m is defined by the formula

(5.6) $$\tilde{d}\omega = d\omega + m\,\pi \wedge \omega ,$$

where π is the Reeb form. We see that $\tilde{d}\omega$ is also half–invariant of weight m as follows. From (5.5), we compute

(5.7) $$d\omega^* = \tau^m d\omega + m\,\tau^m d(\log \tau) \wedge \omega .$$

Then, using (5.4), we have

(5.8) $\tilde{d}\omega^* = d\omega^* + m\,\pi^* \wedge \omega^* = \tau^m\,(d\omega + m\,\pi \wedge \omega) = \tau^m\,\tilde{d}\omega$.

Using the rules for ordinary exterior differentiation, it is straightforward to verify the following rules for the half–invariant differentiation process. Suppose that ω_1 and ω_2 are half–invariant and that ω_1 is a q–form, then

(5.9) $\tilde{d}(\omega_1 \wedge \omega_2) = \tilde{d}\omega_1 \wedge \omega_2 + (-1)^q\,\omega_1 \wedge \tilde{d}\omega_2$.

If ω_1 and ω_2 are of the same weight, then

(5.10) $\tilde{d}(\omega_1 + \omega_2) = \tilde{d}\omega_1 + \tilde{d}\omega_2$.

Unlike ordinary exterior differentiation, differentiating twice by this method does not lead to zero, in general. In fact, if ω has weight m, then

$$\tilde{d}(\tilde{d}\omega) = d(d\omega + m\,\pi \wedge \omega) + m\,\pi \wedge (d\omega + m\,\pi \wedge \omega) .$$

Using (5.9) and the equation $d(d\omega) = 0$, one can reduce the equation above to

(5.11) $\tilde{d}(\tilde{d}\omega) = m\,d\pi \wedge \omega$.

From (5.11), we see that $\tilde{d}(\tilde{d}\omega) = 0$ for every form ω only when $d\pi = 0$. In that case, since we are working locally, we can take U to be contractible, and thus $\pi = d\sigma$, for some smooth scalar function σ on U. Then, if we take a renormalization with $\tau = e^\sigma$, we have from (5.4) that $\pi^* = 0$. Equation (5.6) implies that after this renormalization, half–invariant differentiation is just ordinary exterior differentiation. This special choice of τ is determined up to a constant factor. In cases where $d\pi \neq 0$, there may not be such an optimal choice of renormalization. In those cases, $\tilde{d}\tilde{d} \neq 0$, but there are still certain commutation relations which hold for the computing of second derivatives. These are of primary importance.

For the rest of this section, we adopt the following convention on the indices:

$$1 \le i, j, k \le n-1 \ .$$

Suppose now that f is a half–invariant scalar function of weight m on U. Using the fact that $\tilde{d}f$ has weight m and that the basis forms have weight 1, we can write

(5.14) $\tilde{d}f = f_1 \theta_1 + \dots + f_{n-1} \theta_{n-1}$,

where the f_i are half–invariant functions of weight $m-1$. We call the f_i the *half-invariant derivatives of f with respect to the basis* $\{ \theta_i \}$. We now derive the commutation relations which are satisfied by the half–invariant derivatives of these f_i. Using (5.11), we take the half–invariant derivative of (5.14) and get

(5.15) $m f \, d\pi = \sum \tilde{d}f_i \wedge \theta_i + f_i \tilde{d}\theta_i \ .$

We express the 2–forms $d\pi$ and $\tilde{d}\theta_i$ in terms of the basis forms as follows:

(5.16) $d\pi = \sum_{i<j} p_{ij} \, \theta_i \wedge \theta_j \ , \qquad \tilde{d}\theta_k = \sum_{i<j} c^k_{ij} \, \theta_i \wedge \theta_j \ .$

For convenience in subsequent formulas, we also define the coefficients for the case $i \ge j$, by setting

(5.17) $p_{ji} = - p_{ij} \ , \qquad c^k_{ji} = - c^k_{ij} \ , \quad \text{for all } i, j, k \ .$

By (5.14), we have

(5.18) $\tilde{d}f_i = \sum f_{ij} \, \theta_j \ .$

If we now substitute (5.16) and (5.18) into (5.15), we obtain the following equation,

$$m f \sum_{i<j} p_{ij} \, \theta_i \wedge \theta_j \; = \; \sum_i \sum_j f_{ij} \, \theta_j \wedge \theta_i + \sum_k f_k \, \tilde{d}\theta_k$$

(5.19)

$$= \; \sum_i \sum_j f_{ij} \, \theta_j \wedge \theta_i + \sum_k f_k \, \sum_{i<j} c_{ij}^{\,k} \, \theta_i \wedge \theta_j \, .$$

By taking the coefficient of $\theta_i \wedge \theta_j$ in (5.19), we obtain the *commutation relations*

(5.20) $$f_{ij} - f_{ji} \; = \; \sum_k f_k \, c_{ij}^{\,k} \; - \; m f \, p_{ij} \, .$$

By the convention (5.17), these are valid for all i, j. Thus, the commutation relations for the second derivatives of a half–invariant function f can be determined from the p_{ij}, $c_{ij}^{\,k}$, f and the first derivatives of f. This is crucial in the solution of the differential equations involved in a typical application of this method. The classification of Dupin hypersurfaces in $\mathbb{R}^4$ in the next section is a good example.

4.6 Dupin Hypersurfaces in 4–Space

In this section, we give Pinkall's [1], [3] local classification of proper Dupin hypersurfaces in $\mathbb{R}^4$ up to Lie equivalence (see also Cecil–Chern [2]). The umbilic case with $g = 1$ distinct curvature sphere is well known. Theorem 3.1 handles the case $g = 2$. Thus, the only remaining case is that of $g = 3$. This classification is rather involved, and it makes use of the method of moving frames in a way that was not necessary in the case of $g = 2$. It is the first case where Lie invariants are necessary in the classification, and it is worthy of careful study. The partial classification of Dupin hypersurfaces in $\mathbb{R}^5$ due to Niebergall [1] is similar in spirit to this case.

We follow the notation of the previous section, and consider a proper Dupin submanifold $\lambda : M^3 \to \Lambda^7$ with 3 distinct curvature spheres at each point. Locally, λ is Lie equivalent to a Dupin hypersurface immersed in $\mathbb{R}^4$, but we do not assume that the Euclidean projection of λ into $\mathbb{R}^4$ is an immersion. As before, we make a local choice of Lie frame $\{Y_1,...,Y_7\}$ on an open subset U of M^3 so that for each $x \in M^3$, we have $\lambda(x) = [Y_1(x), Y_7(x)]$. We can take Y_1 and

Y_7 to be curvature spheres at each point of U. Furthermore, by applying Theorem 4.1 first to Y_1 and then to Y_7, we can arrange that the Maurer–Cartan forms for the Lie frame satisfy

(6.1) $\omega_1^3 = 0$, $\omega_1^7 = 0$; $\omega_7^4 = 0$, $\omega_7^1 = 0$.

Next, by making a change of frame of the form

(6.2) $Y_1^* = \sigma Y_1$, $Y_2^* = (1/\sigma)Y_2$, $Y_7^* = \tau Y_7$, $Y_6^* = (1/\tau)Y_6$,

for suitable smooth functions σ and τ on U, we can arrange that $Y_1 + Y_7$ represents the third curvature sphere at each point of U. Then, we can use the method in the proof of Theorem 4.1 to find a Lie frame whose Maurer–Cartan forms satisfy

(6.3) $\omega_1^5 + \omega_7^5 = 0$, $\omega_1^1 - \omega_7^7 = 0$,

as well as (6.1). Conditions (6.1) and (6.3) completely determine Y_3 , Y_4 and Y_5 ,while Y_1 and Y_7 are determined up to a transformation of the form

(6.4) $Y_1^* = \tau Y_1$, $Y_7^* = \tau Y_7$,

for some smooth non–vanishing function τ, i.e., a renormalization.

Each of the three curvature maps Y_1 , Y_7 and $Y_1 + Y_7$ is constant along the leaves of its corresponding principal foliation. Thus, each of these maps factors through an immersion of the corresponding 2–dimensional space of leaves of its principal foliation into Q^5. In terms of moving frames, this implies that the forms ω_1^4 , ω_1^5 and ω_7^3 are linearly independent, i.e.,

(6.5) $\omega_1^4 \wedge \omega_1^5 \wedge \omega_7^3 \neq 0$.

This can also be seen by expressing the forms above in terms of a Lie frame $\{Z_1,...,Z_{n+3}\}$ whose Maurer–Cartan forms satisfy the regularity condition (4.6). For simplicity, we will also use the notation as in Section 4.5,

(6.6) $$\theta_1 = \omega_1^4 , \quad \theta_2 = \omega_1^5 , \quad \theta_3 = \omega_7^3 .$$

Analytically, the Dupin conditions are three partial differential equations, and we are treating an over–determined system. The method of moving frames reduces the handling of its integrability conditions to a straightforward algebraic problem, viz., that of repeated exterior differentiations. Later, we will use the method of half–invariant differentiation to simplify some of the calculations.

We begin by computing the exterior derivatives of the equations $\omega_1^3 = 0$, $\omega_7^4 = 0$ and $\omega_1^5 + \omega_7^5 = 0$. These come from the fact that Y_1 , Y_7 and $Y_1 + Y_7$ are curvature spheres. Using the skew–symmetry of (4.3), as well as the relations (6.1) and (6.3), the exterior derivatives of these three equations yield the system

(6.7)
$$
\begin{aligned}
0 &= \omega_1^4 \wedge \omega_3^4 + \omega_1^5 \wedge \omega_3^5 , \\
0 &= \qquad\quad \omega_1^5 \wedge \omega_4^5 + \omega_7^3 \wedge \omega_3^4 , \\
0 &= \omega_1^4 \wedge \omega_4^5 \qquad\quad + \omega_7^3 \wedge \omega_3^5 .
\end{aligned}
$$

If we take the wedge product of the first of these equations with ω_1^4 , we conclude that ω_3^5 is in the span of ω_1^4 and ω_1^5 . On the other hand, taking the wedge product of the third equation with ω_1^4 yields that ω_3^5 is in the span of ω_1^4 and ω_3^7 . Consequently, $\omega_3^5 = \rho\omega_1^4$, for some smooth function ρ. Similarly, there exist smooth functions α and β such that $\omega_3^4 = \alpha\omega_1^5$ and $\omega_4^5 = \beta\omega_7^3$. Then, if we substitute these results into (6.7), we get that $\rho = \alpha = \beta$, and hence, we have

(6.8) $$\omega_3^5 = \rho\omega_1^4 , \quad \omega_3^4 = \rho\omega_1^5 , \quad \omega_4^5 = \rho\omega_7^3 .$$

This function ρ plays a crucial role in the classification. Next, we differentiate the three equations which come from the Dupin conditions,

(6.9) $$\omega_1^7 = 0 , \quad \omega_7^1 = 0 , \quad \omega_1^1 - \omega_7^7 = 0 .$$

As above, the use of the skew–symmetry relations and (6.1), (6.3) yields the existence of smooth functions $a, b, c, p, q, r, s, t, u$ such that the following relations hold:

(6.10)
$$\omega_4^7 = -\omega_6^4 = a\omega_1^4 + b\omega_1^5 ,$$
$$\omega_5^7 = -\omega_6^5 = b\omega_1^4 + c\omega_1^5 ;$$

(6.11)
$$\omega_3^1 = -\omega_2^3 = p\omega_7^3 - q\omega_1^5 ,$$
$$\omega_5^1 = -\omega_2^5 = q\omega_7^3 - r\omega_1^5 ;$$

(6.12)
$$\omega_4^1 = -\omega_2^4 = b\omega_1^5 + s\omega_1^4 + t\omega_7^3 ,$$
$$\omega_6^3 = -\omega_3^7 = q\omega_1^5 + t\omega_1^4 + u\omega_7^3 .$$

We next take the exterior derivatives of the three basis forms ω_1^4 , ω_1^5 and ω_7^3 . Using the relations that we have derived so far, we obtain from the Maurer–Cartan equation (4.4)

$$d\omega_1^4 = \omega_1^1 \wedge \omega_1^4 + \omega_1^5 \wedge \omega_5^4 = \omega_1^1 \wedge \omega_1^4 - \rho\omega_1^5 \wedge \omega_7^3 .$$

We obtain similar expressions for $d\omega_1^5$ and $d\omega_7^3$ in the same way. Now, when we write these expressions in terms of θ_1 , θ_2 , θ_3 , they become

(6.13)
$$d\theta_1 = \omega_1^1 \wedge \theta_1 - \rho\, \theta_2 \wedge \theta_3 ,$$
$$d\theta_2 = \omega_1^1 \wedge \theta_2 - \rho\, \theta_3 \wedge \theta_1 ,$$
$$d\theta_3 = \omega_1^1 \wedge \theta_3 - \rho\, \theta_1 \wedge \theta_2 .$$

As we noted in Section 4.5, the form $\pi = -\omega_1^1$ can always be used as a Reeb form for half–invariant differentiation. Using the formula

$$\tilde{d}\omega = d\omega + m\,\pi \wedge \omega ,$$

for the half–invariant derivative of a form ω of weight m, and recalling that the θ_i have weight 1, we can rewrite (6.13) as

$$\tilde{d}\theta_1 = -\rho\,\theta_2 \wedge \theta_3\,,$$

(6.14)
$$\tilde{d}\theta_2 = -\rho\,\theta_3 \wedge \theta_1\,,$$

$$\tilde{d}\theta_3 = -\rho\,\theta_1 \wedge \theta_2\,.$$

From (5.16), (5.17) and (6.14), we see that the functions $c_{ij}^{\,k}$ for the half–invariant differentiation have the form

(6.15) $\quad c_{23}^{1} = c_{31}^{2} = c_{12}^{3} = -\rho\,,\quad c_{32}^{1} = c_{13}^{2} = c_{21}^{3} = \rho\,,\quad c_{ij}^{\,k} = 0\,,$ otherwise.

We next differentiate (6.8). We have $\omega_3^{4} = \rho\omega_1^{5}$. On the one hand,

$$d\omega_3^{4} = \rho\,d\omega_1^{5} + d\rho \wedge \omega_1^{5}\,.$$

Using the second equation in (6.13) with $\omega_1^{5} = \theta_2$, this becomes

$$d\omega_3^{4} = \rho\,\omega_1^{1} \wedge \omega_1^{5} - \rho^2\,\omega_7^{3} \wedge \omega_1^{4} + d\rho \wedge \omega_1^{5}\,.$$

On the other hand, we can compute $d\omega_3^{4}$ from the Maurer–Cartan equation (4.4) and use the relationships that we have derived to find

$$d\omega_3^{4} = (-p - \rho^2 - a)(\omega_1^{4} \wedge \omega_7^{3}) - q\,\omega_1^{5} \wedge \omega_1^{4} + b\,\omega_7^{3} \wedge \omega_1^{5}\,.$$

Equating these two expressions for $d\omega_3^{4}$ yields

(6.16) $\quad (-p - a - 2\rho^2)\,\omega_1^{4} \wedge \omega_7^{3} = (d\rho + \rho\omega_1^{1} - q\omega_1^{4} - b\omega_7^{3}) \wedge \omega_1^{5}\,.$

Because of the independence of ω_1^{4}, ω_1^{5} and ω_7^{3}, both sides of the equation above must vanish. Thus, we conclude that

(6.17) $\qquad\qquad\qquad 2\rho^2 = -a - p,$

and that $d\rho + \rho\omega_1^{1} - q\omega_1^{4} - b\omega_7^{3}$ is a multiple of ω_1^{5}. Similarly, differentiation of $\omega_4^{5} = \rho\omega_7^{3}$ yields the following analogue of (6.16),

(6.18) $(s - a - r + 2\rho^2)\, \omega_1^2 \wedge \omega_1^5 = (d\rho + \rho\omega_1^1 + t\omega_1^5 - q\omega_1^4) \wedge \omega_7^3$,

and differentiation of $\omega_3^5 = \rho\omega_1^4$ yields

(6.19) $(c + p + u - 2\rho^2)\, \omega_1^5 \wedge \omega_7^3 = (- d\rho - \rho\, \omega_1^1 - t\, \omega_1^5 + b\, \omega_7^3) \wedge \omega_1^4$.

In each of the equations (6.16), (6.18), (6.19), both sides of the equation must vanish. From the vanishing of the left sides of the equations, we get the fundamental relationship,

(6.20) $2\rho^2 = - a - p = a + r - s = c + p + u$.

Furthermore, from the vanishing of the right sides of the three equations (6.16), (6.18), (6.19), we can determine after some algebra that

(6.21) $d\rho + \rho\, \omega_1^1 = q\, \omega_1^4 - t\, \omega_1^5 + b\, \omega_7^3$.

This equation shows the importance of ρ. From (6.8), we see that the half–invariant function ρ has weight -1, since ω_3^5 has weight 0 and ω_1^4 has weight 1. Recalling that $\pi = - \omega_1^1$, we have

(6.22) $d\rho + \rho\, \omega_1^1 = d\rho - \rho\, \pi = \tilde{d}\rho = \rho_1\, \theta_1 + \rho_2\, \theta_2 + \rho_3\, \theta_3$.

Comparing (6.21) and (6.22), we find that the half–invariant derivatives ρ_i are given by

(6.23) $\rho_1 = q$, $\rho_2 = - t$, $\rho_3 = b$.

Using the Maurer–Cartan equations, we can compute

$$
\begin{aligned}
d\omega_1^1 &= \omega_1^4 \wedge \omega_4^1 + \omega_1^5 \wedge \omega_5^1 \\
&= \omega_1^4 \wedge (b\, \omega_1^5 + t\, \omega_7^3) + \omega_1^5 \wedge (q\, \omega_7^3 - r\, \omega_1^5) \\
&= b\, \omega_1^4 \wedge \omega_1^5 + q\, \omega_1^5 \wedge \omega_7^3 - t\, \omega_7^3 \wedge \omega_1^4 .
\end{aligned}
$$

Using (6.6) and (6.23), this can be rewritten as

$$(6.24) \qquad d\omega_1^1 = \rho_3\, \theta_1 \wedge \theta_2 + \rho_1\, \theta_2 \wedge \theta_3 + \rho_2\, \theta_3 \wedge \theta_1 \ .$$

Since $\pi = - \omega_1^1$, comparing (6.24) and (5.16) yields the following formulas for the coefficients p_{ij} of the half–invariant differentiation,

$$(6.25) \qquad p_{21} = - p_{12} = \rho_3 \ , \quad p_{32} = - p_{23} = \rho_1 \ , \quad p_{13} = - p_{31} = \rho_2 \ .$$

By (6.15) and (6.25), the fundamental coefficients for the half–invariant differentiation process are all expressed in terms of ρ and its half–invariant derivatives ρ_i. This enables us to express the half–invariant derivatives of all the functions involved in terms of ρ and its successive half–invariant derivatives. Ultimately, this leads to the solution of the problem.

First note that (6.15) and (6.25) allow us to write the commutation relations (5.20) for a half–invariant function σ of weight m in the form

$$(6.26) \qquad \begin{aligned} \sigma_{12} - \sigma_{21} &= - \rho\sigma_3 + m\sigma\rho_3 \ , \\ \sigma_{23} - \sigma_{32} &= - \rho\sigma_1 + m\sigma\rho_1 \ , \\ \sigma_{31} - \sigma_{13} &= - \rho\sigma_2 + m\sigma\rho_2 \ . \end{aligned}$$

In particular, since ρ is half–invariant of weight $- 1$, we have

$$(6.27) \qquad \begin{aligned} \rho_{12} - \rho_{21} &= - 2\,\rho\,\rho_3 \ , \\ \rho_{23} - \rho_{32} &= - 2\,\rho\,\rho_1 \ , \\ \rho_{31} - \rho_{13} &= - 2\,\rho\,\rho_2 \ . \end{aligned}$$

We next take the exterior derivative of equations (6.10) – (6.12). We first differentiate the equation

$$(6.28) \qquad \omega_4^7 = a\, \omega_1^4 + b\, \omega_1^5 \ .$$

On the one hand, from the Maurer–Cartan equation (4.4) for $d\omega_4^7$, and omitting those terms which are already known to vanish, we have

$$
\begin{aligned}
d\omega_4^7 &= \omega_4^2 \wedge \omega_2^7 + \omega_4^3 \wedge \omega_3^7 + \omega_4^5 \wedge \omega_5^7 + \omega_4^7 \wedge \omega_7^7 \\
&= -\omega_1^4 \wedge \omega_2^7 + (-\rho\,\omega_1^7) \wedge (-q\,\omega_1^5 - t\,\omega_1^4 - u\,\omega_7^3) \\
&\quad + \rho\,\omega_7^3 \wedge (b\,\omega_1^4 + c\,\omega_1^5) + (a\,\omega_1^4 + b\,\omega_1^5) \wedge \omega_1^1 .
\end{aligned}
$$
(6.29)

On the other hand, differentiation of the right side of (6.28) yields

$$
\begin{aligned}
d\omega_4^7 &= da \wedge \omega_1^4 + a\,d\omega_1^4 + db \wedge \omega_1^5 + b\,d\omega_1^5 \\
&= da \wedge \omega_1^4 + a\,(\omega_1^1 \wedge \omega_1^4 - \rho\,\omega_1^5 \wedge \omega_7^3) \\
&\quad + db \wedge \omega_1^5 + b\,(\omega_1^1 \wedge \omega_1^5 - \rho\,\omega_1^4 \wedge \omega_7^3) .
\end{aligned}
$$
(6.30)

Equating (6.29) and (6.30), we find

(6.31)
$$
\begin{aligned}
&(da + 2a\,\omega_1^1 - 2b\,\rho\,\omega_7^3 - \omega_2^7) \wedge \omega_1^4 \\
&+ (db + 2b\,\omega_1^1 + (a + u - c)\,\rho\,\omega_7^3) \wedge \omega_1^5 + \rho\,t\,\omega_1^4 \wedge \omega_1^5 = 0 .
\end{aligned}
$$

Since $b = \rho_3$ is half–invariant of weight -2, we have

(6.32) $db + 2b\,\omega_1^1 = d\rho_3 + 2\,\rho_3\,\omega_1^1 = \tilde{d}\rho_3 = \rho_{31}\theta_1 + \rho_{32}\theta_2 + \rho_{33}\theta_3 .$

By examining the coefficient of $\omega_1^5 \wedge \omega_7^3 = \theta_2 \wedge \theta_3$ in (6.31) and using (6.32), we find

(6.33) $\rho_{33} = \rho\,(c - a - u) .$

Furthermore, the remaining terms in (6.31) are

(6.34)
$$
\begin{aligned}
&(da + 2a\,\omega_1^1 - \omega_2^7 - 2\,\rho b\,\omega_7^3 - (\rho t + \rho_{31})\,\omega_1^5) \wedge \omega_1^4 \\
&+ \text{terms involving } \omega_1^5 \text{ and } \omega_7^3 \text{ only.}
\end{aligned}
$$

Thus, the coefficient in parentheses must be a multiple of ω_1^4, call it $\bar{a}\,\omega_1^4$. Since $\omega_4^7 = \,< dY_4, Y_6 >$ has weight -1 and ω_1^4 has weight 1, it follows from (6.10) that the function a has weight -2. Using (6.23), we can write (6.34) as

(6.35) $da + 2a\,\omega_1^1 = da - 2a\pi = \tilde{d}a = \omega_2^7 + \bar{a}\,\theta_1 + (\rho_{31} - \rho\,\rho_2)\,\theta_2 + 2\rho\,\rho_3\,\theta_3 .$

In a similar manner, if we differentiate $\omega_5^7 = b\,\omega_1^4 + c\,\omega_1^5$, and use the fact that the function c has weight -2, we obtain

$$(6.36) \quad dc + 2c\,\omega_1^1 = dc - 2c\pi = \tilde{dc} = \omega_2^7 + (\rho_{32} + \rho\,\rho_1)\,\theta_1 + \bar{c}\,\theta_2 - 2\rho\,\rho_3\,\theta_3\,.$$

Thus, from the two equations in (6.10), we have obtained (6.33), (6.35), and (6.36). In a completely analogous manner, we can differentiate the two equations in (6.11) to obtain

$$(6.37) \qquad\qquad \rho_{11} = \rho\,(s + r - p)\,,$$

$$(6.38) \quad dp + 2p\,\omega_1^1 = \tilde{dp} = -\omega_2^7 + 2\rho\,\rho_1\,\theta_1 + (-\rho_{13} - \rho\,\rho_2)\,\theta_2 + \bar{p}\,\theta_3\,,$$

$$(6.39) \quad dr + 2r\,\omega_1^1 = \tilde{dr} = -\omega_2^7 - 2\rho\,\rho_1\,\theta_1 + \bar{r}\,\theta_2 + (-\rho_{12} + \rho\,\rho_3)\,\theta_3\,.$$

Similarly, differentiation of (6.12) yields

$$(6.40) \qquad\qquad \rho_{22} + \rho_{33} = \rho\,(p - r - s)\,,$$

$$(6.41) \quad ds + 2s\,\omega_1^1 = \tilde{ds} = \bar{s}\,\theta_1 + (\rho_{31} + \rho\,\rho_2)\,\theta_2 + (-\rho_{21} + \rho\,\rho_3)\,\theta_3\,,$$

$$(6.42) \quad du + 2u\,\omega_1^1 = \tilde{du} = (-\rho_{23} - \rho\,\rho_1)\,\theta_1 + (\rho_{13} - \rho\,\rho_2)\,\theta_2 + \bar{u}\,\theta_3\,.$$

In these equations, the coefficients $\bar{a}$, $\bar{c}$, $\bar{p}$, $\bar{r}$, $\bar{s}$ and $\bar{u}$ remain undetermined. However, by differentiating (6.20) and using the appropriate equations from above, one can show that

$$(6.43) \quad
\begin{array}{ll}
\bar{a} = -6\rho\,\rho_1\,, & \bar{c} = 6\rho\,\rho_2\,, \\
\bar{p} = -6\rho\,\rho_3\,, & \bar{r} = 6\rho\,\rho_2\,, \\
\bar{s} = -12\rho\,\rho_1\,, & \bar{u} = 12\rho\,\rho_3\,.
\end{array}$$

From equations (6.33), (6.37), (6.40) and (6.20), we can easily compute that

$$(6.44) \qquad\qquad \rho_{11} + \rho_{22} + \rho_{33} = 0\,.$$

Using (6.43), equations (6.41) and (6.42) can be rewritten as

(6.45) $\quad ds + 2s\, \omega_1^1 = \tilde{ds} = -12\rho\, \rho_1\, \theta_1 + (\rho_{31} + \rho\, \rho_2)\, \theta_2 + (-\rho_{21} + \rho\, \rho_3)\, \theta_3$,

(6.46) $\quad du + 2u\, \omega_1^1 = \tilde{du} = (-\rho_{23} - \rho\, \rho_1)\, \theta_1 + (\rho_{13} - \rho\, \rho_2)\, \theta_2 + 12\, \rho\, \rho_3\, \theta_3$.

Thus, the half–invariant derivatives of s and u are expressed in terms of ρ and its half–invariant derivatives. By taking half–invariant derivatives of these two equations and making use of (6.44) and the commutation relations (6.26) for ρ and its various derivatives, one can show after a lengthy calculation that the following fundamental equations hold:

(6.47)
$$\rho\, \rho_{12} + \rho_1\, \rho_2 + \rho^2\, \rho_3 = 0 ,$$
$$\rho\, \rho_{21} + \rho_1\, \rho_2 - \rho^2\, \rho_3 = 0 ,$$
$$\rho\, \rho_{23} + \rho_2\, \rho_3 + \rho^2\, \rho_1 = 0 ,$$
$$\rho\, \rho_{32} + \rho_2\, \rho_3 - \rho^2\, \rho_1 = 0 ,$$
$$\rho\, \rho_{31} + \rho_3\, \rho_1 + \rho^2\, \rho_2 = 0 ,$$
$$\rho\, \rho_{13} + \rho_3\, \rho_1 - \rho^2\, \rho_2 = 0 .$$

We now briefly outline the details of this calculation. By (6.45), we have

(6.48) $\qquad s_1 = -12\, \rho\, \rho_1 , \quad s_2 = \rho_{31} + \rho\, \rho_2 , \quad s_3 = \rho\, \rho_3 - \rho_{21}$.

The half–invariant quantity s has weight -2, so the commutation relation (6.26) gives

(6.49) $\qquad s_{12} - s_{21} = -2s\, \rho_3 - \rho\, s_3 = -2s\, \rho_3 - \rho\, (\rho\, \rho_3 - \rho_{21})$.

On the other hand, by taking half–invariant derivatives of (6.48), we can compute directly that

(6.50) $\qquad s_{12} - s_{21} = -12\rho\, \rho_{12} - 12\, \rho_2\, \rho_1 - (\rho_{311} + \rho_1\, \rho_2 + \rho\, \rho_{21})$.

The main problem now is to get the half–invariant derivative ρ_{311} into a workable form. By taking the half–invariant derivative of the third equation in (6.27), we find

$$(6.51) \qquad \rho_{311} - \rho_{131} = -2\,\rho_1\,\rho_2 - 2\rho\,\rho_{21}\,.$$

Then, using the commutation relation,

$$\rho_{131} = \rho_{113} - 2\,\rho_1\,\rho_2 - \rho\,\rho_{12}\,,$$

we get from (6.51),

$$(6.52) \qquad \rho_{311} = \rho_{113} - 4\,\rho_1\,\rho_2 - \rho\,\rho_{12} - 2\rho\,\rho_{21}\,.$$

Taking the half–invariant derivative of $\rho_{11} = \rho\,(s + r - p)$, and substituting the expression obtained for ρ_{113} into (6.52), we get

$$(6.53) \qquad \rho_{311} = \rho_3\,(s + r - p) - 3\rho\,\rho_{21} - 2\rho\,\rho_{12} + 8\rho^2\,\rho_3 - 4\rho_1\,\rho_2\,.$$

If we substitute (6.53) for ρ_{311} in (6.50) and then equate (6.49) and (6.50), we obtain the first equation in (6.47). The cyclic permutations are obtained in a similar way from $s_{23} - s_{32}$, etc.

Although Y_3 , Y_4 and Y_5 are already completely determined by the conditions (6.1) and (6.3), it is still possible to make a change of frame of the form

$$(6.54) \qquad
\begin{aligned}
Y_1^* &= \tau\, Y_1\,, & Y_2^* &= (1/\tau)\,Y_2 + \mu\,Y_7\,, \\
Y_7^* &= \tau\, Y_7\,, & Y_6^* &= (1/\tau)\,Y_6 - \mu\,Y_1\,.
\end{aligned}$$

Under this change, we have

$$(6.55) \qquad
\begin{aligned}
\omega_1^{4*} &= \tau\,\omega_1^4\,, & \omega_1^{5*} &= \tau\,\omega_1^5\,, & \omega_7^{3*} &= \tau\,\omega_7^3\,, \\
\omega_4^{7*} &= (1/\tau)\,\omega_4^7 + \mu\,\omega_1^4\,, \\
\omega_3^{1*} &= (1/\tau)\,\omega_3^1 - \mu\,\omega_7^3\,.
\end{aligned}$$

Suppose that we write

$$\omega_4^{7*} = a^* \, \omega_1^{4*} + b^* \, \omega_1^{5*} \, ,$$
$$\omega_3^{1*} = p^* \, \omega_7^{3*} - q^* \, \omega_1^{5*} \, .$$

Then from (6.55), we obtain

$$a^* = \tau^{-2} a + \tau^{-1} \mu \, ,$$
$$p^* = \tau^{-2} p - \tau^{-1} \mu \, .$$

Thus, by taking $\mu = (p - a) / 2\tau$, we can arrange that $a^* = p^*$. We now make this change of frame and drop the asterisks. In this new frame, we have

(6.56) $a = p = -\rho^2, \quad r = 3\rho^2 + s, \quad c = 3\rho^2 - \mu \, .$

Using the fact that $a = p$, we can subtract (6.38) from (6.35) and get that

(6.57) $\omega_2^7 = 4\rho \, \rho_1 \, \theta_1 - ((\rho_{31} + \rho_{13}) / 2) \, \theta_2 - 4\rho \, \rho_3 \, \theta_3 \, .$

Now through (6.35) – (6.39), the half–invariant derivatives of the functions a, c, p and r are expressed in terms of ρ and its derivatives. We are now ready to proceed to the main results. Ultimately, we show that it is possible to choose a frame in which ρ is constant. Thus, the classification naturally splits into two cases, $\rho = 0$ and $\rho \neq 0$. We handle the two cases separately.

Case 1: $\rho \neq 0$.

Assume now that the function ρ is never zero on M^{n-1}. The key step in getting ρ to be constant is the following lemma due to Pinkall [3, p.108], where his function c is the negative of our function ρ. The proof here was first given in Cecil–Chern [2, p.33], and it is somewhat simpler than Pinkall's. The crucial point is that since $\rho \neq 0$, the fundamental equations (6.47) allow us to express all of the second half–invariant derivatives ρ_{ij} of ρ in terms of ρ and its first derivatives.

Lemma 6.1: *Suppose that the function ρ never vanishes on M^{n-1}. Then, its half–invariant derivatives satisfy $\rho_1 = \rho_2 = \rho_3 = 0$ at every point of M^{n-1}.*

Proof: First, note that if the function ρ_3 vanishes identically, then (6.47) and the assumption that $\rho \neq 0$ imply that ρ_1 and ρ_2 also vanish identically. We now complete the proof of the lemma by showing that ρ_3 must vanish everywhere. This is accomplished by considering the expression $s_{12} - s_{21}$. By the commutation relations (6.26), we have

$$s_{12} - s_{21} = -2s\,\rho_3 - \rho\,s_3\,.$$

By (6.47) − (6.48), we see that

$$\rho\,s_3 = \rho^2\,\rho_3 - \rho\,\rho_{21} = \rho_1\,\rho_2\,,$$

and so

(6.58) $$s_{12} - s_{21} = -2s\,\rho_3 - \rho_1\,\rho_2\,.$$

On the other hand, we can compute s_{12} by taking the derivative of the equation $s_1 = -12\rho\,\rho_1$. Then, using the expression for ρ_{12} obtained from (6.47), we get

(6.59) $$\begin{aligned} s_{12} &= -12\,\rho_2\,\rho_1 - 12\rho\,\rho_{12} = -12(\rho_2\,\rho_1 + \rho\,\rho_{12}) \\ &= -12(\rho_2\,\rho_1 + (-\rho_2\,\rho_1 - \rho^2\rho_3)) = 12\,\rho^2\,\rho_3\,. \end{aligned}$$

Next, we have from (6.48) that $s_2 = \rho_{31} + \rho\,\rho_2$. Using (6.47), we have

$$\rho_{31} = -\rho_3\,\rho_1\,\rho^{-1} - \rho\,\rho_2\,,$$

and thus,

(6.60) $$s_2 = -\rho_3\,\rho_1\,/\,\rho\,.$$

Then, we compute

$$s_{21} = -\left(\rho\left(\rho_3\,\rho_{11} + \rho_{31}\,\rho_1\right) - \rho_3\,\rho_1^2\right) / \rho^2 \, .$$

Using (6.37) for ρ_{11} and (6.47) to get ρ_{31}, this becomes

(6.61) $$s_{21} = -\rho_3\,(s + r - p) + 2\,\rho_3\,\rho_1^2\,\rho^{-2} + \rho_1\,\rho_2 \, .$$

Now, equate the expression (6.58) for $s_{12} - s_{21}$ with the expression obtained by subtracting (6.61) from (6.59) to get

$$-2s\,\rho_3 - \rho_1\,\rho_2 = 12\,\rho^2\,\rho_3 + \rho_3\,(s + r - p) - 2\,\rho_3\,\rho_1^2\,\rho^{-2} - \rho_1\,\rho_2 \, .$$

This can be rewritten as

(6.62) $$0 = \rho_3\,(12\,\rho^2 + 3s + r - p - 2\rho_1^2\,\rho^{-2}) \, .$$

Using the expressions in (6.56) for r and p, we see that $3s + r - p = 4s + 4\rho^2$, and so, (6.62) can be written as

(6.63) $$0 = \rho_3\,(16\,\rho^2 + 4s - 2\,\rho_1^2\,\rho^{-2}) \, .$$

Suppose that $\rho_3 \neq 0$ at some point $x \in M^{n-1}$. Then ρ_3 does not vanish on some neighborhood U of x. By (6.63), we have

(6.64) $$16\,\rho^2 + 4s - 2\,\rho_1^2\,\rho^{-2} = 0$$

on U. We now take the θ_2–half–invariant derivative of (6.64) and obtain

(6.65) $$32\rho\,\rho_2 + 4s_2 - 4\rho_1\,\rho_{12}\,\rho^{-2} + 4\rho_1^2\,\rho_2\,\rho^{-3} = 0 \, .$$

We can now substitute the expression (6.60) for s_2 and the formula

$$\rho_{12} = -\rho_1\,\rho_2\,\rho^{-1} - \rho\,\rho_3$$

obtained from (6.47) into (6.65). After some algebra, (6.65) reduces to

$$\rho_2 \, (32 \, \rho^4 + 8 \, \rho_1^2) = 0 \, .$$

Since $\rho \neq 0$, this implies that $\rho_2 = 0$ on U. But then, the left side of the equation below, obtained from (6.47),

$$\rho \, \rho_{21} + \rho_1 \, \rho_2 = \rho^2 \, \rho_3 \, ,$$

must vanish on U. Since $\rho \neq 0$, we conclude that $\rho_3 = 0$ on U, a contradiction to our assumption. Hence, ρ_3 must vanish identically on M^{n-1}, and the lemma is proven. □

We now continue with the case $\rho \neq 0$. According to the previous lemma, all the half–invariant derivatives of ρ are zero, and our formulas simplify greatly. Equations (6.33) and (6.37) give

$$c - a - u = 0 \, , \qquad s + r - p = 0 \, .$$

These combined with (6.56) give

(6.66) $$c = r = \rho^2, \qquad u = -s = 2 \, \rho^2 \, .$$

By (6.57), we have $\omega_2^7 = 0$. So, the differentials of the frame vectors can now be written

(6.67)
$$
\begin{aligned}
dY_1 - \omega_1^1 \, Y_1 &= \omega_1^4 \, Y_4 + \omega_1^5 \, Y_5 \, , \\
dY_7 - \omega_1^1 \, Y_7 &= \omega_7^3 \, Y_3 - \omega_1^5 \, Y_5 \, , \\
dY_2 + \omega_1^1 \, Y_2 &= \rho^2 (\omega_7^2 \, Y_3 + 2 \, \omega_1^3 \, Y_4 + \omega_1^5 \, Y_5) \, , \\
dY_6 + \omega_1^1 \, Y_6 &= \rho^2 (2 \, \omega_7^2 \, Y_3 + \omega_1^3 \, Y_4 - \omega_1^5 \, Y_5) \, , \\
dY_3 &= \omega_7^3 \, Z_3 + \rho \, (\omega_1^5 \, Y_4 + \omega_1^4 \, Y_5) \, , \\
dY_4 &= - \omega_1^4 \, Z_4 + \rho \, (- \omega_1^5 \, Y_3 + \omega_7^3 \, Y_5) \, , \\
dY_5 &= \omega_1^5 \, Z_5 + \rho \, (- \omega_1^4 \, Y_3 - \omega_7^3 \, Y_4) \, ,
\end{aligned}
$$

where

$$Z_3 = -Y_6 + \rho^2 (-Y_1 - 2Y_7),$$

(6.68) $$Z_4 = Y_2 + \rho^2 (2Y_1 + Y_7),$$

$$Z_5 = -Y_2 + Y_6 + \rho^2 (-Y_1 + Y_7).$$

From this, we notice that

(6.69) $$Z_3 + Z_4 + Z_5 = 0,$$

so that the points Z_3, Z_4 and Z_5 lie on a line. From (6.22), (6.24) and the lemma above, we see that

(6.70) $$\tilde{d}\rho = d\rho + \rho\,\omega_1^1 = d\rho - \rho\pi = 0, \quad d\omega_1^1 = -d\pi = 0.$$

Here we have the special situation $d\pi = 0$, which we discussed after equation (5.11) in Section 4.5. In this case, there is always a renormalization, determined up to a constant factor, which makes $\pi^* = 0$. From (6.70), we have $\pi = d(\log \rho)$. Hence, we take the renormalization factor $\tau = e^{\log \rho} = \rho$. This makes $\pi^* = \omega_1^{1*} = 0$. Furthermore, since the function ρ, as defined in (6.8), has weight -1, we also get that $\rho^* = \rho^{-1}\rho = 1$ in the new frame. Thus, we now make a change of frame of the form

(6.71) $$Y_1^* = \rho\, Y_1, \quad Y_7^* = \rho\, Y_7, \quad Y_2^* = (1/\rho)\, Y_2, \quad Y_6^* = (1/\rho)\, Y_6,$$

$$Y_i^* = Y_i, \quad i = 3, 4, 5.$$

We see from (6.68) that the Z_i are half–invariant of weight -1. Further, the basis forms ω_1^4, ω_1^5 and ω_7^3 are of weight 1. Thus, we have

(6.72) $$Z_i^* = (1/\rho)\, Z_i, \quad i = 3, 4, 5,$$

$$\omega_1^{4*} = \rho\, \omega_1^4, \quad \omega_1^{5*} = \rho\, \omega_1^5, \quad \omega_7^{3*} = \rho\, \omega_7^3.$$

Using the equations above, we compute the differentials of the frame vectors as follows:

(6.73)
$$dY_1^* = \omega_1^{4*} Y_4 + \omega_1^{5*} Y_5 ,$$
$$dY_7^* = \omega_7^{3*} Y_3 - \omega_1^{5*} Y_5 ,$$
$$dY_2^* = \omega_7^{3*} Y_3 + 2\,\omega_1^{4*} Y_4 + \omega_1^{5*} Y_5 ,$$
$$dY_6^* = 2\,\omega_7^{3*} Y_3 + \omega_1^{4*} Y_4 - \omega_1^{5*} Y_5 ,$$
$$dY_3 = \omega_7^{3*} Z_3^* + \omega_1^{5*} Y_4 + \omega_1^{4*} Y_5 ,$$
$$dY_4 = -\omega_1^{4*} Z_4^* - \omega_1^{5*} Y_3 + \omega_7^{3*} Y_5 ,$$
$$dY_5 = \omega_1^{5*} Z_5^* - \omega_1^{4*} Y_3 - \omega_7^{3*} Y_4 ,$$

with

(6.74)
$$dZ_3^* = 2(\, -2\omega_7^{3*} Y_3 - \omega_1^{4*} Y_4 + \omega_1^{5*} Y_5 \,) ,$$
$$dZ_4^* = 2(\, \omega_7^{3*} Y_3 + 2\,\omega_1^{4} Y_4 + \omega_1^{5*} Y_5 \,) ,$$
$$dZ_5^* = 2(\, \omega_7^{3*} Y_3 - \omega_1^{4*} Y_4 - 2\,\omega_1^{5*} Y_5 \,) ,$$

and

(6.75)
$$d\omega_1^{4*} = -\omega_1^{5*} \wedge \omega_7^{3*} , \quad \text{i.e.,} \quad d\theta_1^* = -\theta_2^* \wedge \theta_3^* ,$$
$$d\omega_1^{5*} = -\omega_7^{3*} \wedge \omega_1^{4*} , \quad \text{i.e.,} \quad d\theta_2^* = -\theta_3^* \wedge \theta_1^* ,$$
$$d\omega_7^{3*} = -\omega_1^{4*} \wedge \omega_1^{5*} , \quad \text{i.e.,} \quad d\theta_3^* = -\theta_1^* \wedge \theta_2^* .$$

Comparing the last equation with (6.13), we see that $\omega_1^{1*} = 0$ and $\rho^* = 1$. This is the final frame needed in our case of $\rho \neq 0$, so we drop the asterisks once more.

We can now prove Pinkall's classification for the case $\rho \neq 0$. As with the cyclides of Dupin, there is only one compact model up to Lie equivalence. This is Cartan's isoparametric hypersurface in S^4. It is a tube of constant radius over each of its two focal submanifolds in S^4, both of which are Veronese surfaces. (See Cecil–Ryan [7, pp.296–299] for more detail.)

Theorem 6.1: (a) *Every connected Dupin submanifold* $\lambda : M^3 \to \Lambda^7$ *with* $\rho \neq 0$ *is contained in a unique compact, connected Dupin submanifold with* $\rho \neq 0$.
(b) *Any two Dupin submanifolds with* $\rho \neq 0$ *are locally Lie equivalent, each being Lie equivalent to an open subset of an isoparametric hypersurface in* S^4.

Proof: Let $\{Y_a\}$ be the frame just constructed, whose derivatives satisfy (6.73). The three curvature curvature spheres of M^3 are Y_1, Y_7 and $Y_1 + Y_7$.

Let

$$W_1 = -Y_1 + Y_6 - 2Y_7, \quad W_2 = -2Y_1 + Y_2 - Y_7.$$

Then from (6.73), we find that $dW_1 = dW_2 = 0$. Hence, W_1 and W_2 are constant. Furthermore, since

$$< W_1, W_1 > = < W_2, W_2 > = -4, \quad < W_1, W_2 > = -2,$$

the line $[W_1, W_2]$ is timelike. Finally, the equations

$$< Y_1, W_1 > = 0, \quad < Y_7, W_2 > = 0, \quad < Y_1 + Y_7, W_1 - W_2 > = 0,$$

imply that λ is Lie equivalent to an open subset of an isoparametric hypersurface by Theorem 5.6 of Chapter 3. Since any open subset of an isoparametric hypersurface is contained in a unique compact isoparametric hypersurface (see Münzner [1] or Cecil–Ryan [7, Sections 4–6 of Chapter 3]), part (a) is proven. Furthermore, because all isoparametric hypersurfaces in S^4 with 3 principal curvatures are locally Lie equivalent by a result of Cartan [3], part (b) is also true.   □

Case 2: $\rho = 0$.

We now consider the case where ρ is identically zero. It turns out that these are all reducible to cyclides in $\mathbb{R}^3$. We return to the frame that we used prior to the assumption that $\rho \neq 0$. Thus, only those relations through equation (6.57) are valid. Since ρ is identically zero, so are all of its half–invariant derivatives. From (6.23) and (6.56), we see that the functions defined in (6.10) – (6.12) satisfy

$$q = t = b = 0, \quad a = p = 0, \quad r = s, \quad c = -u.$$

Thus, from (6.57) we have $\omega_2^7 = 0$. From these and the other relations among the Maurer–Cartan forms which we have derived, we see that the differentials of the frame vectors can be written

$$
\begin{aligned}
dY_1 - \omega_1^1\, Y_1 &= \omega_1^4\, Y_4 + \omega_1^5\, Y_5 \,, \\
dY_7 - \omega_1^1\, Y_7 &= \omega_7^3\, Y_3 - \omega_1^5\, Y_5 \,, \\
dY_2 + \omega_1^1\, Y_2 &= s\,(-\omega_1^4\, Y_4 + \omega_1^5\, Y_5) \,, \\
dY_6 + \omega_1^1\, Y_6 &= u\,(\omega_7^3\, Y_3 + \omega_1^5\, Y_5) \,, \\
dY_3 &= \omega_7^3\,(-Y_6 + u\, Y_7) \,, \\
dY_4 &= \omega_1^4\,(s\, Y_1 - Y_2) \,, \\
dY_5 &= \omega_1^5\,(-s\, Y_1 - Y_2 + Y_6 - u\, Y_7) \,.
\end{aligned}
$$

(6.76)

Note also that from (6.45) and (6.46), we have

(6.77) $$\tilde{d}s = ds + 2s\,\omega_1^1 = 0 \,, \qquad \tilde{d}u = du + 2u\,\omega_1^1 = 0 \,.$$

From (6.24), we have $d\omega_1^1 = 0$ and thus, $d\pi = 0$. Again, we follow the procedure given after (5.11) to find a renormalization in which $\pi^* = 0$. Specifically, we assume that we are working in a contractible local neighborhood U, so that $\pi = d\sigma$ for some smooth scalar function σ on U. We then take a renormalization (5.1) with $\tau = e^\sigma$. This results in $\pi^* = -\omega_1^{1*} = 0$, and (6.76) can be rewritten as

$$
\begin{aligned}
dY_1^* &= \omega_1^{4*}\, Y_4 + \omega_1^{5*}\, Y_5 \,, \\
dY_7^* &= \omega_7^{3*}\, Y_3 - \omega_1^{5*}\, Y_5 \,, \\
dY_2^* &= s^*\,(-\omega_1^{4*}\, Y_4 + \omega_1^{5*}\, Y_5) \,, \\
dY_6^* &= u^*\,(\omega_7^{3*}\, Y_3 + \omega_1^{5*}\, Y_5) \,, \\
dY_3^* &= \omega_7^{3*}\, Z_3^* \,, \quad \text{where } Z_3^* = -Y_6^* - u^*\, Y_7^* \,, \\
dY_4^* &= \omega_1^{4*}\, Z_4^* \,, \quad \text{where } Z_4^* = s^*\, Y_1^* - Y_2^* \,, \\
dY_5^* &= \omega_1^{5*}\, Z_5^* \,, \quad \text{where } Z_5^* = -s^*\, Y_1^* - Y_2^* + Y_6^* - u^*\, Y_7^* \,,
\end{aligned}
$$

(6.78)

where

(6.79) $$s^* = \tau^{-2} s \,, \qquad u^* = \tau^{-2} u \,.$$

After this renormalization, half–invariant differentiation is the same as ordinary exterior differentiation. Thus, from (6.77) and (6.79) we have

$$(6.80) \qquad\qquad ds^* = 0 , \qquad du^* = 0 ,$$

i.e., s^* and u^* are constant functions on the local neighborhood U.

The frame (6.78) is our final frame, and we will drop the asterisks once more. Since the functions s and u are now constant, we can compute from (6.78) that

$$(6.81) \qquad \begin{aligned} dZ_3 &= -2u\, \omega_7^3\, Y_3 , \\ dZ_4 &= 2s\, \omega_1^4\, Y_4 , \\ dZ_5 &= 2(u-s)\, \omega_1^5\, Y_5 . \end{aligned}$$

From this we see that the following 4–dimensional subspaces of $\mathbb{P}^6$,

$$(6.82) \qquad \begin{aligned} &\text{Span } \{Y_1 , Y_4 , Y_5 , Z_4 , Z_5\} , \\ &\text{Span } \{Y_7 , Y_3 , Y_5 , Z_3 , Z_5\} , \\ &\text{Span } \{Y_1 + Y_7 , Y_3 , Y_4 , Z_3 , Z_4\} , \end{aligned}$$

are invariant under exterior differentiation, and hence, they are constant. Thus, each of the three curvature sphere maps Y_1, Y_7 and $Y_1 + Y_7$ is contained in a 4–dimensional subspace of $\mathbb{P}^6$. One can easily show that each of the spaces in (6.82) has signature $(4, 1)$. Thus, by Theorem 2.8, our Dupin submanifold is Lie equivalent to a tube over a cyclide in $\mathbb{R}^3$ in three different ways. Hence, we have the following result due to Pinkall.

Theorem 6.2: *Any Dupin submanifold $\lambda : M^3 \to \Lambda^7$ with $\rho = 0$ is reducible. It is locally Lie equivalent to a tube over a cyclide of Dupin in $\mathbb{R}^3 \subset \mathbb{R}^4$.*

Pinkall [3, p.111] proceeds to classify Dupin submanifolds with $\rho = 0$ up to Lie equivalence. We will not do that here. The reader can follow Pinkall's proof using the fact that his constants α and β are our constants s and $-u$, respectively.

References

Abresch, U.
[1] *Isoparametric hypersurfaces with four or six distinct principal curvatures*, Math. Ann. 264 (1983), 283–302.

Arnold, V. I.
[1] *Mathematical methods of classical mechanics*, Springer, Berlin, etc., 1978.

Artin, E.
[1] *Geometric algebra*, Wiley–Interscience, New York, 1957.

Banchoff, T.
[1] *The spherical two-piece property and tight surfaces in spheres*, J. Diff. Geom. 4 (1970), 193–205.

Blair, D.
[1] *Contact manifolds in Riemannian geometry*, Lecture Notes in Math. 509, Springer, Berlin, etc., 1976.

Blaschke, W.
[1] *Vorlesungen über Differentialgeometrie und geometrische Grundlagen von Einsteins Relativitätstheorie*, Vol. 3, Springer, Berlin, etc., 1929.

Bol, G.

[1] *Projektive Differentialgeometrie*, Vol. 3, Vandenhoeck and Ruprecht, Göttingen, 1967.

Buyske, S.

[1] *Lie sphere transformations and the focal sets of certain taut immersions*, Trans. Amer. Math. Soc. 311 (1989), 117–133.

Cahen, M. and Kerbrat, Y.

[1] *Domaines symétriques des quadriques projectives*, J. Math. Pures et Appl. 62 (1983), 327–348.

Cartan, E.

[1] *Théorie des groupes finis et continus et la géométrie différentielle traitées par la méthode du repère mobile*, Gauthiers–Villars, Paris, 1937.

[2] *Familles de surfaces isoparamétriques dans les espaces à courbure constante*, Annali di Mat. 17 (1938), 177–191.

[3] *Sur des familles remarquables d'hypersurfaces isoparamétriques dans les espaces sphériques*, Math. Z. 45 (1939), 335–367.

[4] *Sur quelques familles remarquables d'hypersurfaces*, C.R. Congrès Math. Liège, 1939, 30–41.

[5] *Sur des familles d'hypersurfaces isoparamétriques des espaces sphériques à 5 et à 9 dimensions*, Revista Univ. Tucuman, Serie A, 1 (1940), 5–22.

[6] *The theory of spinors*, Hermann, Paris, 1966, reprinted by Dover, New York, 1981.

Carter, S. and West, A.,

[1] *Tight and taut immersions*, Proc. London Math. Soc. (3), 25 (1972), 701–720.

[2] *Isoparametric systems and transnormality*, Proc. London Math. Soc. (3), 51 (1985), 520–542.

[3] *Generalized Cartan polynomials*, J. London Math. Soc. (2), 32 (1985), 305–316.

[4]   *Isoparametric and totally focal submanifolds*, Proc. London Math. Soc.
       (3) 60 (1990), 609–624.

Cecil, T.
[1]   *Taut immersions of non-compact surfaces into a Euclidean 3-space*,
       J. Diff. Geom. 11 (1976), 451–459.
[2]   *Reducible Dupin submanifolds*, Geometriae Dedicata 32 (1989),
       281–300.
[3]   *On the Lie curvatures of Dupin hypersurfaces*, Kodai Math. J. 13 (1990),
       143–153.
[4]   *Lie sphere geometry and Dupin submanifolds*, Geometry and Topology
       of Submanifolds, III, ed. L. Verstraelen and A. West, 90–107, World
       Scientific, Singapore, etc., 1991.

Cecil, T. and Chern, S.-S.
[1]   *Tautness and Lie sphere geometry*, Math. Ann. 278 (1987), 381–399.
[2]   *Dupin submanifolds in Lie sphere geometry*, Differential Geometry and
       Topology, Proceedings Tianjin 1986–87, Lecture Notes in Math. 1369,
       1–48, Springer, Berlin, etc., 1989.

Cecil, T. and Ryan, P.
[1]   *Focal sets of submanifolds*, Pacific J. Math. 78 (1978), 27–39.
[2]   *Focal sets, taut embeddings and the cyclides of Dupin*, Math. Ann. 236
       (1978), 177–190.
[3]   *Distance functions and umbilic submanifolds of hyperbolic space*,
       Nagoya Math. J. 74 (1979), 67–75.
[4]   *Tight and taut immersions into hyperbolic space*, J. London Math. Soc.
       (2), 19 (1979), 561–572.
[5]   *Conformal geometry and the cyclides of Dupin*, Canadian J. Math. 32
       (1980), 767–782.
[6]   *Tight spherical embeddings*, Global Differential Geometry and Global
       Analysis, Proceedings Berlin 1979, Lecture Notes in Math. 838, 94–104,
       Springer, Berlin, etc., 1981.

[7] *Tight and taut immersions of manifolds*, Research Notes in Math. 107, Pitman, London, 1985.

Dorfmeister, J. and Neher, E.

[1] *Isoparametric hypersurfaces, case $g = 6$, $m = 1$*, Communications in Algebra 13 (1985), 2299–2368.

Dupin, C.

[1] *Applications de géométrie et de méchanique*, Paris, 1822.

Ferus, D., Karcher, H. and Münzner, H.

[1] *Cliffordalgebren und neue isoparametrische Hyperflächen*, Math. Z. 177 (1981), 479–502.

Fillmore, J.

[1] *The fifteen-parameter conformal group*, International J. Theoretical Physics 16 (1977), 937–963.

[2] *On Lie's higher sphere geometry*, L'Enseignement Math. 25 (1979), 77–114.

[3] *Hermitean quadrics as contact manifolds*, Rocky Mountain J. Math. 14 (1984), 559–571.

Griffiths, P.

[1] *On Cartan's method of Lie groups and moving frames as applied to uniqueness and existence questions in differential geometry*, Duke Math. J. 41 (1974), 775–814.

Grove, K. and Halperin S.

[1] *Dupin hypersurfaces, group actions, and the double mapping cylinder*, J. Diff. Geom. 26 (1987), 429–457.

Harle, C.

[1] *Isoparametric families of submanifolds*, Bol. Soc. Bras. Mat. 13 (1982), 35–48.

Hawkins, T.
[1] *Line geometry, differential equations and the birth of Lie's theory of
     groups*, The History of Modern Mathematics, Vol. I, 275–327, Academic
     Press, San Deigo, 1989.

Hebda, J.
[1] *The possible cohomology of certain types of taut submanifolds*, Nagoya
     Math. J. 111 (1988), 85–97.

Heintze, E., Olmos, C. and Thorbergsson, G.
[1] *Submanifolds with constant principal curvatures and normal holonomy
     groups*, to appear in International J. Math.

Hsiang, W.–Y., Palais, R. and Terng, C.–L.
[1] *The topology of isoparametric submanifolds*, J. Diff. Geom. 27 (1988),
     423–460.

Jensen, G.
[1] *Higher order contact of submanifolds of homogeneous spaces*, Lecture
     Notes in Math. 610, Springer, Berlin, etc., 1977.

Klein, F.
[1] *Vorlesungen über höhere Geometrie*, Springer, Berlin 1926, 3 Aufl.,
     reprinted by Chelsea, New York, 1957.

Kobayashi, S. and Nomizu, K.
[1] *Foundations of differential geometry*, I, II, Wiley–Interscience, New
     York, 1963 and 1969.

Lie, S.
[1] *Über Komplexe, inbesondere Linien- und Kugelkomplexe, mit
     Anwendung auf der Theorie der partieller Differentialgleichungen*, Math.
     Ann. 5 (1872), 145–208, 209–256 (Ges. Abh. 2, 1–121).

Lie, S. and Scheffers, G.

[1] *Geometrie der Berührungstransformationen*, Teubner, Leipzig, 1896, reprinted by Chelsea, New York, 1977.

Lilienthal, R.

[1] *Besondere Flächen*, Encyklopädie der Math. Wissenschaften, Vol. III, 3, 269–354, Teubner, Leipzig, 1902–1927.

Milnor, J.

[1] *Morse theory*, Ann. Math. Stud. 51, Princeton U. Press, 1963.

Miyaoka, R.

[1] *Compact Dupin hypersurfaces with three principal curvatures*, Math. Z. 187 (1984), 433–452.

[2] *Dupin hypersurfaces and a Lie invariant*, Kodai Math. J. 12 (1989), 228–256.

[3] *Dupin hypersurfaces with six principal curvatures*, Kodai Math. J. 12 (1989), 308–315.

[4] *Lie contact structures and normal Cartan connections*, Kodai Math. J. 14 (1991), 13–41.

[5] *Lie contact structures and conformal structures*, Kodai Math. J. 14 (1991), 42–71.

Miyaoka, R. and Ozawa, T.

[1] *Construction of taut embeddings and Cecil-Ryan conjecture*, Geometry of Manifolds, ed. K. Shiohama, 181–189, Academic Press, New York, 1989.

Münzner, H.–F.

[1] *Isoparametrische Hyperflächen in Sphären*, I and II, Math Ann. 251 (1980), 57–71 and 256 (1981), 215–232.

Niebergall, R.

[1] *Dupin hypersurfaces in $\mathbb{R}^5$*, I and II, to appear in Geometriae Dedicata.

Nomizu, K.
[1]   *Fundamentals of linear algebra*, McGraw–Hill, New York, 1966.
[2]   *Characteristic roots and vectors of a differentiable family of symmetric matrices*, Lin. and Multilin. Alg. 2 (1973), 159–162.
[3]   *Some results in E. Cartan's theory of isoparametric families of hypersurfaces*, Bull. Amer. Math. Soc. 79 (1973), 1184–1188.
[4]   *Elie Cartan's work on isoparametric families of hypersurfaces*, Proc. Symposia in Pure Math., Amer. Math. Soc. 27 (Part I), (1975), 191–200.

O'Neill, B.
[1]   *Semi-Riemannian geometry*, Academic Press, New York, 1983.

Ozawa, T.
[1]   *On the critical sets of distance functions to a taut submanifold*, Math. Ann. 276 (1986), 91–96.

Ozeki, H. and Takeuchi, M.
[1]   *On some types of isoparametric hypersurfaces in spheres*, I and II, Tôhoku Math. J. 27 (1975), 515–559 and 28 (1976), 7–55.

Palais, R. and Terng, C.–L.
[1]   *A general theory of canonical forms*, Trans. Amer. Math. Soc. 300 (1987), 771–789.
[2]   *Critical point theory and submanifold geometry*, Lecture Notes in Math. 1353, Springer, Berlin, etc., 1988.

Pinkall, U.
[1]   *Dupin'sche Hyperflächen*, Dissertation, Univ. Freiburg, 1981.
[2]   *W-Kurven in der ebenen Lie-Geometrie*, Elemente der Mathematik 39 (1984), 28–33 and 67–78.
[3]   *Dupin'sche Hyperflächen in E^4*, Manuscr. Math. 51 (1985), 89–119.
[4]   *Dupin hypersurfaces*, Math. Ann. 270 (1985), 427–440.

[5] *Curvature properties of taut submanifolds*, Geometriae Dedicata 20 (1986), 79–83.

Pinkall, U. and Thorbergsson, G.

[1] *Deformations of Dupin hypersurfaces*, Proc. Amer. Math. Soc. 107 (1989), 1037–1043.

[2] *Examples of infinite dimensional isoparametric submanifolds*, Math. Z. 205 (1990), 279–286.

Reckziegel, H.

[1] *Krümmungsflächen von isometrischen Immersionen in Räume konstanter Krümmung*, Math. Ann. 223 (1976), 169–181.

[2] *On the eigenvalues of the shape operator of an isometric immersion into a space of constant curvature*, Math. Ann. 243 (1979), 71–82.

[3] *Completeness of curvature surfaces of an isometric immersion*, J. Diff. Geom. 14 (1979), 7–20.

Rowe, D,

[1] *The early geometrical works of Sophus Lie and Felix Klein*, The History of Modern Mathematics, Vol. I, 209–273, Academic Press, San Deigo, 1989.

Ryan, P.

[1] *Homogeneity and some curvature conditions for hypersurfaces*, Tôhoku Math. J. 21 (1969), 363–388.

[2] *Euclidean and non-Euclidean geometry*, Cambridge Univ. Press, Cambridge, 1986.

Samuel, P.

[1] *Projective geometry*, Springer, Berlin, etc., 1988.

Sasaki, S. and Suguri, T.

[1] *On the problems of equivalence of plane curves in the Lie's higher circle geometry and of minimal curves in the conformal geometry*, Tôhoku Math. J. 47 (1940), 77–86.

Singley, D.

[1] *Smoothness theorems for the principal curvatures and principal vectors of a hypersurface*, Rocky Mountain J. Math. 5 (1975), 135–144.

Solomon, B.

[1] *The harmonic analysis of cubic isoparametric minimal hypersurfaces* I: *dimensions 3 and 6*, Amer. J. Math. 112 (1990), 157–203.

Spivak, M.

[1] *A comprehensive introduction to differential geometry*, 1–5, Publish or Perish, Boston, 1970–1975.

Sternberg, S.

[1] *Lectures on differential geometry*, Prentice–Hall, Englewood Cliffs, 1964, second edition, Chelsea, New York, 1983.

Strübing, W.

[1] *Isoparametric submanifolds*, Geometriae Dedicata 20 (1986), 367–387.

Takagi, R. and Takahashi, T.

[1] *On the principal curvatures of homogeneous hypersurfaces in a sphere*, Differential Geometry in honor of K. Yano, Kinokuniya, Tokyo, 1972, 469–481.

Terng, C.–L.

[1] *Isoparametric submanifolds and their Coxeter groups*, J. Diff. Geom. 21 (1985), 79–107.

[2] *Convexity theorem for isoparametric submanifolds*, Invent. Math. 85 (1986), 487–492.

[3]   *Submanifolds with flat normal bundle*, Math. Ann. 277 (1987), 95–111.

[4]   *Proper Fredholm submanifolds of Hilbert space*, J. Diff. Geom. 29 (1989), 9–47.

[5]   *Recent progress in submanifold geometry*, to appear in Proc. Symposia in Pure Math. (Differential Geometry, 1990), Amer. Math. Soc.

Thorbergsson, G.

[1]   *Dupin hypersurfaces*, Bull. London Math. Soc. 15 (1983), 493–498.

[2]   *Isoparametric foliations and their buildings*, Ann. Math. 133 (1991), 429–446.

Voss, K.

[1]   *Eine Verallgemeinerung der Dupinschen Zykliden*, Tagungsbericht 41/1981, Geometrie, Mathematisches Forschungsinstitut, Oberwolfach, 1981.

Wang, Q.–M.

[1]   *Isoparametric functions on Riemannian manifolds. I*, Math. Ann. 277 (1987), 639–646.

[2]   *On the topology of Clifford isoparametric hypersurfaces*, J. Diff. Geom. 27 (1988), 55–66.

Index

Universitext *(continued)*